Computational Statistics

An Introduction to R

Computational Statistics

Günther Sawitzki

CRC Press is an imprint of the
Taylor & Francis Group an **informa** business
A CHAPMAN & HALL BOOK

Chapman & Hall/CRC
Taylor & Francis Group
6000 Broken Sound Parkway NW, Suite 300
Boca Raton, FL 33487-2742

Printed in the United States of America on acid-free paper
10 9 8 7 6 5 4 3 2 1

International Standard Book Number-13: 978-1-4200-8678-2 (Hardcover)

Library of Congress Cataloging-in-Publication Data

Sawitzki, Günther.
Computational statistics : an introduction to R / Günther Sawitzki.
p. cm.
Includes bibliographical references and index.
ISBN 978-1-4200-8678-2 (hardcover : alk. paper)
1. R (Computer program language) 2. Mathematical statistics--Data processing. I. Title.

QA276.45.R3S29 2009
519.50285--dc22 2008042298

Visit the Taylor & Francis Web site at
http://www.taylorandfrancis.com

and the CRC Press Web site at
http://www.crcpress.com

Introduction

This introduction to R is intended as course material to be used in a concise course or for self-instruction. The course is for students with basic knowledge of stochastics. Notions like distribution function, quantile, expected value and variance are presupposed, as well as some familiarity with statistical features corresponding to these notions. Only a condensed summary is included here. Classical distributions, such as binomial, uniform, and Gaussian should be familiar, together with derived distributions and their asymptotic behaviour. An in-depth knowledge of statistics is not required. However, while this course does cover much of statistics, it is not a substitute for a general statistics course. This course concentrates on the "computing" aspects. Statistical concepts and statistical points of view are introduced and discussed. For an in-depth discussion, the reader is referred to statistics courses.

A working knowledge in computer usage is presupposed, at least rudimentary knowledge of programming concepts like variables, loops, functions. No extended knowledge in computing is required.

What Is R?

R is a programming language, and the name of a software system that implements this language [31]. The R programming language has been developed for statistics and stochastic simulation. By now, it has become a standard in these fields. To be precise, we should use specific terms: the language is called S, its implementation and the environment are called R. The original authors of S are John M. Chambers, R. A. Becker and A. R. Wilks, AT&T Bell Laboratories, Statistics Research Department. The language and its development are documented in a series of books, commonly referred to according to the colour of their cover as white ([5]), blue ([2]) and green book ([4]).

For a long time the AT&T implementation of S has been the reference for the S language. Today, S is available in a commercial version called S-Plus `<http://www.insightful.com/>` (based on the AT&T implementation) and as a free software version R[1], or "Gnu S" `<http://www.r-project.org/>`.

[1] R got its name by accident — the same accident that made the first names of the original authors of R (Ross Ihaka and Robert Gentleman) start with R.

In the meantime, R has become the reference implementation. Essential more precise definitions and — if necessary — even modifications of the language are given by R. For simplicity, here and in the sequel we use "R language" as a common term even where the precise term should be "the S language using the R implementation".

R is an interpreted programming language. Instructions in R are executed immediately. In addition to the original elements of S, R has several extensions. Some of these have been introduced in response to recent developments in statistics, some are introduced to open experimental facilities. Advancements in the S language are taken into account.

The most recent (2008) version of R is 2.x. This version is largely compatible with the previous version R 1.x. The essential changes are internal to the system. For initial use, there is no significant difference from R, 1.x. For the advanced user, there are three essential innovations:

- *Graphics:* The basic graphics system in R implements a model inspired by pen and paper drawing. A graphics port (paper) is opened, and lines, points or symbols are drawn in this port. In R 2.x there is a second additional graphics system, oriented at a viewport/object model. Graphical objects in various positions and orientations are mapped in a visual space.
- *Packages:* The original R had a linear command history and a uniform workspace. R 2.x introduced an improved support for "packages" that may have encapsulated information. To support this, language concepts such as "name spaces" and various tools have been introduced.
- *Internationalisation:* Originally, R was based on the English language, and ASCII was the general encoding used. With R 2.x extensive support for other languages and encodings has been introduced. With this, it has become possible to provide localised versions.

Two aspects of R are active areas of recent developments: interactive access and integration in a networked environment. These and other aspects are part of Omegahat, an attempt to develop a next generation system based on the experiences from R. This more experimental project is accessible at `<http://www.omegahat.org/>`. R does already provide simple possibilities to call functions implemented in other languages like C or FORTRAN. Omegahat extends these possibilities and allows direct access to Java, Perl .... A Java-based graphical interface for R is JGR, accessible at `<http://stats.math.uni-augsburg.de/software/>`. A collection of interactive displays for R is in *`iplots`*, available at the same site.

Recent developments related to R are in `<http://r-forge.r-project.org/>`. Many helpful extensions are in `<http://www.bioconductor.org/>`, a site that is targeted at biocomputing.

R has been developed for practical work in statistics. Usability often has been given priority over abstract design principles. As a consequence, it is not easy to give a systematic introduction to R. A winding path is chosen here instead: case studies and examples, followed by systematic surveys. Practical work should make use of the rich online material that comes with R. Starting points are the "frequently asked questions" (FAQ) `<http://www.cran.r-project.org/faqs.html>`. "An Introduction to R" ([30]) is the "official" introduction. This documentation and other manuals can be downloaded from `<http://www.cran.r-project.org/manuals.html>`.

R comes with an extensive online help system providing function descriptions and examples. Once you have become familiar with R, the online help and manuals will become your first source of information. For off-line reference, we include some examples of information provided by the help system, with kind permission of The R Foundation for Statistical Computing.

Many R functions are included in the basic system. Other functions must be loaded from libraries. Some of these libraries are distributed with the R system. If additional packages are needed, `<http://www.cran.r-project.org/>` is a first source for downloads, and `<http://r-forge.r-project.org/>` for current projects.

The commercial S-Plus is available in various versions. S-Plus 4.x and S-Plus 2000 use S version 3 and are largely compatible with R. S-Plus 5 is an implementation of S version 4 with some changes that need attention. These programming details of S-Plus are not discussed here. Information about S-Plus can be found at `<http://www.insightful.com/>`.

Contents and Outline of This Course

In its basic version, R contains more than 1500 functions, too many to introduce in just one course, and too many to learn. This course can only open the door to R.

Course participants can come from various backgrounds, with different prerequisites. For pupils and younger students, a mere programming course on technical basics may be appropriate. Later, questions about meaningful classification and background will be more important. This is the aim of the present course. The "technical" material provides a skeleton. Beyond this, we try to open the view for statistical questions and to stimulate interest in the background. This course should whet the appetite for the substance that may be offered in a subsequent well-founded statistical course.

The first part of this course material is organised by themes, using example topics to illustrate how R can be used to tackle statistical problems.

The appendix provides a collection of R language elements and functions. During the course, it can serve as a quick reference and perhaps as a starting point and orientation to access the rich information material that comes bundled with R. After the course, it may serve as a note pad. Finally, in the long run for practical work, the online help and online manuals for R are the prime sources of information. This appendix is not meant to be comprehensive. If a concise syntax description or example could be given, it is included. In other cases, the online help information should be consulted.

Using a selection of the exercises, the course can be completed within about four days of work. Conceptually, it is an introduction to statistics with the following topic areas:

- One-sample analysis and distributions
- Regression
- Two-sample problems and comparison of distributions
- Multivariate analysis

A generous time slot for exercises is recommended (an additional half day for the introductory exercises, an additional half day for one of the project exercises). The course can then be covered in one week, provided follow-up facilities are established to answer questions that have come up, and possibilities are available to follow the interest in the statistical background that may have resulted.

At a more leisurely pace, Chapter 1 with its exercises can be used on its own. This should provide a working base to use R, and more material from the subsequent chapters can be added later as needed. The first chapter is fairly selfcontained, including the necessary basic definitions of statistical terms. The other chapters assume that the reader can look up terms if necessary.

Using the course during a term in a weekly class requires more time, since repetitions must be calculated in. Each of the first four chapters will cover about four lectures, plus time for exercises. For this time schedule, a course covering the statistical background is recommended, running in parallel with this one.

For a subsequent self-paced study that goes into detail on R as a programming language, the recommended reading is ([51]).

Statistical literature is evolving, and new publications will be available at the time you read this text. Instead of giving a long list of the relevant literature available at the time this text is written, the sections include keywords that can be used to locate up-to-date literature.

For economic reasons, most of the illustrations are printed in black and white. Colour versions are available at the web site accompanying this book:

`http://sintro.r-forge.r-project.org/.`

Additional material and updates will be available at this site.

CHAPTER 1

Basic Data Analysis

1.1 R Programming Conventions

Like any programming language, R has certain conventions. Here are the basic rules.

R *Conventions*	

Numbers	A point is used as a decimal separator. Numbers can be written in exponential form; the exponential part is introduced by `E`. Numbers can be complex numbers; the imaginary part is marked by `i`.
	Example: `1`
	`2.3`
	`3.4E5`
	`6i+7.8`
	Numbers can take the values `Inf`, `-Inf`, `NaN` for "not a number" and `NA` for "not available" = missing.
	Example: `1/0` results in `Inf`
	`0/0` results in `NaN`
	`NA` is used as a placeholder for missing numbers.

Strings	Strings are delimited by `"` or `'`.
	Example: `"ABC"`
	`'def'`
	`"gh'ij"`

Comments	Comments start with `#` and go to the end of the current line.

To allow for non-trivial examples, we anticipate a detail: in R, `a:b` is a sequence of numbers. If $a \leq b$, `a:b` is the sequence starting at `a` to at most `b` in steps of 1. If $a < b$, `a:b` is the sequence starting at `a` to at least `b` in steps of -1.

R ***Conventions***	
Objects	The basic elements in R are objects. Objects have types, for example `logical` or `integer`. Objects can have a class attribute specifying more complex type information. *Example:* The basis objects in R are vectors.

Names	R objects can have names, by which they can be accessed. Names begin with a letter or a dot, followed by a sequence of letters, digits, or the special characters _ or .
	Examples: `x`
	`y_1`
	Lower- and uppercase are treated as different.
	Examples: `Y87`
	`y87`

Assignments	Assignments have the form *Syntax:* `name <- value` or alternatively `name = value`. *Example:* `a <- 10` `x <- 1:10`
Queries	If only the name of an object is entered, the value of the object is returned. *Example:* `x`
Indices	Vector components are accessed by index. The lowest index is 1. *Example:* `x[3]` The indices can be specified directly, or using symbolic names or rules. *Examples:* `a[1]` the first element `x[-3]` all elements except the third `x[ x^2 < 10]` all elements where $x^2 < 10$

Typographical Conventions

Examples and input code are formatted so that they can be used with "Cut & Paste" and entered as program input. To allow this, punctuation marks are omitted and the input code is shown without a "prompt". For example:

***Example* 0.1:**

Input
```
1 + 2
```
Output
```
3
```

may correspond to a screen output of

```
> 1+2
[1] 3
>
```

Depending on the configuration, the prompt ">" may be represented by a different symbol.

Acknowledgements

Thanks to the R core team for comments and hints. Special thanks to Friedrich Leisch (R core team) and Antony Unwin (Univ. Augsburg) who worked through an early version of this manuscript. Thanks to Rudy Beran, Lucien Birgé, Dianne Cook, Lutz Dümbgen, Jan Johannes, Deepayan Sarkar, Bill Venables, Ali Ünlü and Adalbert Wilhelm for comments and hints.

Thanks to Dagmar Neubauer and Shashi Kumar for helping with the TeX pre-production, and very special thanks to Gerben Wierda for making the necessary tools accessible.

Literature and Additional References

[30] R Development Core Team (2000–2008): An Introduction to R.
See: <http://www.r-project.org/manuals.html>.

[34] R Development Core Team (2000–2008): R Reference Manual.
See: <http://www.r-project.org/manuals.html>.

The Omega Development Group (2008): Omega.
See: <http://www.omegahat.org/>.

[2] Becker, R.A.; Chambers, J.M.; Wilks, A.R. (1988): *The New S Language.*
Chapman & Hall, New York.

[5] Chambers, J.M.; Hastie, T.J. (eds.) (1992): *Statistical Models in S.*
Chapman & Hall, New York.

[6] Cleveland, W.F. (1993): *Visualizing Data.*
Hobart Press, Summit NJ.

[52] Venables, W.N.; Ripley, B.D. (2002): *Modern Applied Statistics with S.*
Springer, Heidelberg.
See: <http://www.stats.ox.ac.uk/pub/MASS4/>.

[51] Venables, W.N.; Ripley, B.D. (2000): *Programming in S.*
Springer, Heidelberg.
See: <http://www.stats.ox.ac.uk/pub/MASS3/Sprog>.

Contents

Help and Inspection	

Help	Documentation and additional information about an object can be requested using `help`.
	Syntax: `help(name)`
	Examples: `help(exp)`
	`help(x)`
	Alternative form `?name`
	Examples: `?exp`
	`?x`
	A hypertext (currently HTML) version of R's online documentation is activated by `help.start()`. This allows us to search by topics, and provides a more structured access to information.

Inspection	`help()` can only provide information that has been prepared in advance. `str()` can inspect the actual state of an object and display this information. *Syntax:* `str(object, ...)` *Examples:* `str(x)`

R ***Conventions***	
Functions	Function calls in R have the form: *Syntax:* `name(argument ... )` *Example:* `e_10 <- exp(10)` This convention holds even when there are no arguments at all. *Example:* To quit R, you call a "quit" function `q()`. Function arguments are treated in a very flexible way. They can have default values, which are used if no explicit argument value is given. *Examples:* `log(x, base = exp(1))`

(cont.)→

R ***Conventions*** (cont.)	
	Functions can be ***polymorphic***. For a polymorphic function, the actual function is determined by the class of the actual arguments. *Examples:* `plot(x)` # a one-dimensional serial plot `plot(x, x^2)` # a two-dimensional scatter plot `summary(x)`
Operators	When applied to vectors, operators operate on each of the vector components. *Example:* For vectors `y, z`, the product `y*z` is the vector of component-wise products. Operators are special functions. They can be called in prefix form (function form). *Example:* `"+"(x, y)` When applied to two operands with different lengths, the smaller operand is repeated cyclically. *Example:* `(1:2)*(1:6)`

1.2 Generation of Random Numbers and Patterns

Our subject is statistical methods. As a first step, we apply the methods in simulations, that is, we use synthetic data. Generating these data is largely under our control. This gives us the opportunity to gain experience with the methods and allows a critical evaluation. Only then will we use the methods for data analysis.

1.2.1 Random Numbers

Random variables with a uniform distribution can be generated by the function `runif()` Using `help(runif)` or `?runif` we get information on how to use this function:

help(runif)

`Uniform` *The Uniform Distribution*

Description

These functions provide information about the uniform distribution on the interval from `min` to `max`. `dunif` gives the density, `punif` gives the distribution function `qunif` gives the quantile function and `runif` generates random deviates.

Usage

```
dunif(x, min=0, max=1, log = FALSE)
punif(q, min=0, max=1, lower.tail = TRUE, log.p = FALSE)
qunif(p, min=0, max=1, lower.tail = TRUE, log.p = FALSE)
runif(n, min=0, max=1)
```

Arguments

`x,q`	vector of quantiles.
`p`	vector of probabilities.
`n`	number of observations. If `length(n) > 1`, the length is taken to be the number required.
`min,max`	lower and upper limits of the distribution. Must be finite.
`log, log.p`	logical; if TRUE, probabilities p are given as log(p).
`lower.tail`	logical; if TRUE (default), probabilities are $P[X \leq x]$, otherwise, $P[X > x]$.

Details

If `min` or `max` are not specified they assume the default values of `0` and `1` respectively. The uniform distribution has density

$$f(x) = \frac{1}{max - min}$$

for $min \leq x \leq max$.
For the case of $u := min == max$, the limit case of $X \equiv u$ is assumed, although there is no density in that case and `dunif` will return `NaN` (the error condition).
`runif` will not generate either of the extreme values unless `max = min` or `max-min` is small compared to `min`, and in particular not for the default arguments.

References

Becker, R. A., Chambers, J. M. and Wilks, A. R. (1988) *The New S Language*. Wadsworth & Brooks/Cole.

See Also

`.Random.seed` about random number generation, `rnorm`, etc for other distributions.

Examples

```
u <- runif(20)

## The following relations always hold :
punif(u) == u
dunif(u) == 1

var(runif(10000))#- ~ = 1/12 = .08333
```

The help information tells us that as an argument for `runif()` we have to supply the number `n` of random variates to generate. As additional arguments for `runif()` we can specify the minimum and the maximum for the range of the random numbers. If we do not specify additional arguments, the default values `min = 0` and `max = 1` are taken. For example, `runif(100)` generates a vector with 100 uniform random numbers with range $(0, 1)$. Calling `runif(100, -10, 10)` generates a vector with 100 uniform random numbers in the range $(-10, 10)$.

The additional arguments can be supplied in the defined order, or specified by name. If the name of the argument is given, the position can be chosen freely. So instead of `runif(100, -10, 10)` it is possible to use `runif(100, min = -10, max = 10)` or `runif(100, max = 10, min = -10)`. Using the name, it is also possible to set only chosen arguments. For example, if the minimum is not specified, the default value for the minimum is taken: using `runif(100, max = 10)` is equivalent with `runif(100, min = 0, max = 10)`. For better readability, we often write the names of arguments, even if it is not necessary.

Each execution of `runif()` generates 100 new uniform random numbers. We can store these numbers.

```
x <- runif(100)
```

generates a new vector of random numbers and stores it in the variable `x`.

```
x
```

returns the values. By default, it is written to the output, and we can inspect the result. We get a graphical representation, the ***serial plot***, a scatterplot of the entries `x` against its running index, by using

```
plot(x)
```

Example 1.1: A Simple Plot

Input

```
x <- runif(100)
plot(x)
```

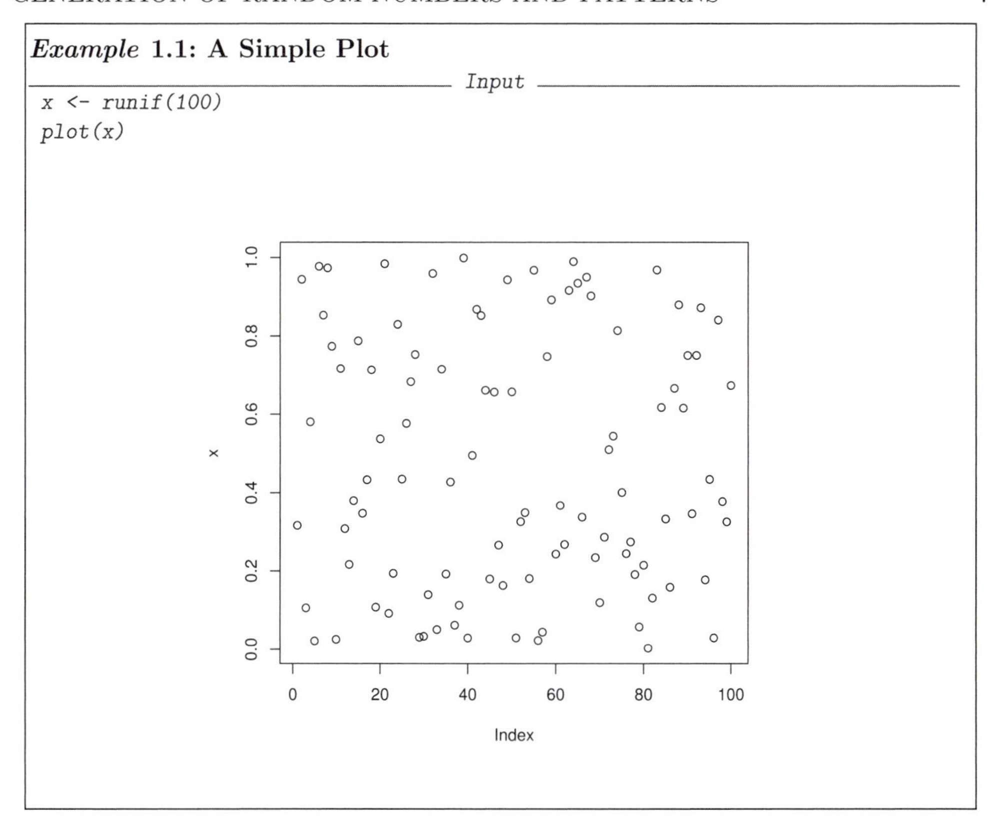

Exercise 1.1	
	Try experimenting with these plots and ***runif()***. Do the plots show images of random numbers? To be more precise: do you accept these plots as images of 100 independent realisations of random numbers, distributed uniformly on $(0, 1)$? Repeat your experiments and try to note as precisely as possible the arguments you have for or against (uniform) randomness. What is your conclusion? Walk through your arguments and try to draft a test strategy to analyse a sequence of numbers for (uniform) randomness. Try to formulate your strategy as clearly as possible. (cont.)→

Exercise 1.1	(cont.)
	Hint: For comparison, you can keep several plots in a window. The code `par(mfrow = c(2, 3))` parametrises the graphics system to show six plots simultaneously, arranged rowwise as a 2×3 matrix (2 rows, 3 columns). The function `par` is the central function to control graphics parameters. For more information, see `help(par)`.

We reveal the secret[1]: the numbers are not random, but completely deterministic. In the background, `runif()` builds a deterministic sequence z_i using iterated functions. In the simplest case, for linear congruence generators, subsequent values z_i, z_{i+1} are generated simply by a linear function. To keep the values in a controlled range, calculation is done modulo an upper bound, that is

$$z_{i+1} = a\ z_i + b \mod M.$$

The resulting values are re-scaled by

$$\frac{z_i}{M} \cdot (\max - \min) + \min$$

and returned to us. To start, an initial value z_0, the ***seed***, must be given. Computationally there is just one variable, `.Random.seed`, which holds the current state z_i for $i = 0, 1, \ldots$, and is managed by the system. The successive values returned are in the hands of the user.

The sequence so defined can be regular and soon lead to a periodic solution. With appropriate choice of the parameters, however, like in the example given in the footnote, it can result in a long period (in the order of magnitude of M) and appear random. But the sequence of numbers is not at all a sequence of independent random numbers, and its distribution is not a uniform distribution on (min, max).

This is the simplest case. Various other algorithms for random number generation are available, but all follow the same scheme. For more information on available random number generators, see `help(.Random.seed)`. See also Appendix A.21 (page A-228).

Even knowing the secret, a lot of additional knowledge is needed to prove that the generated sequence does not follow the rules that apply to a sequence of independent identically distributed random numbers from a uniform distribution. Sequences that claim to work like random numbers are called ***pseudo-random numbers***, if it is important to mark the difference. We use these pseudo-random numbers to generate convenient test data sets. Using these test data sets we can analyse how statistical methods perform

[1] ... at least part of it. The random number generators used in R can be customised and are usually more complex than the simple variant introduced here. For our discussion, the family of linear congruence generators suffices as an illustration. The usual reference here is the "minimal standard generator" with $x_{i+1} = (x_i \times 7^5) \mod 2^{31} - 1$.

under known conditions. In this context, we use pseudo-random numbers as if we had true random numbers.

On the other hand, we can take pseudo-random numbers as a challenge: are we capable of detecting that they are not independent random numbers? If we can detect the difference, we would try to replace the pseudo-random number generator by a better one. But first we have a challenge. Are we capable at all of detecting that the sequence generated by a linear congruence generator, say, is not random but deterministic? If we cannot, what are the intellectual consequences to draw from this?

1.2.2 Patterns

Besides the possibility of generating pseudo-random numbers, R provides several possibilities of generating regular sequences. In many cases these can replace loops, which are common in other languages. Here is an initial survey:

R ***Sequences***	
:	generates a sequence from ⟨`begin`⟩ to at most ⟨`end`⟩. *Syntax:* ⟨`begin`⟩`:`⟨`end`⟩ *Examples:* `1:10` `10.1:1.2`
`c()`	"combine". combines arguments to a new vector. *Syntax:* `c(..., recursive = FALSE)` *Examples:* `c(1, 2, 3)` `c(x, y)` If the arguments are complex data types, the function will descend recursively if called with `recursive = TRUE`.
`seq()`	generates general sequences. *Syntax:* See `help(seq)`
`rep()`	repeats an argument. *Syntax:* `rep(x, times, ...)` *Examples:* `rep(x, 3)` `rep(1:3, c(2, 3, 1))`

Here "..." denotes a variable list of arguments. We will use this notation frequently.

Exercise 1.2	
	Use `plot(sin(1:100))` to generate a plot of a discretised sine function. Use your strategy from Exercise 1.1. Does your strategy detect that the sine function is not a random sequence? *Hint:* If you do not recognise the sine function at first sight, use `plot(sin (1:100), type = "l")` to connect the points.

Listing the numbers of a data set, as for example the output of a random number generator, rarely helps detect underlying structures. Simple unspecific graphical representations like the serial plot have some, but only limited, use. Even with clear patterns the information provided by these plots is rarely meaningful. Purposeful representations are needed to investigate distribution properties of data. The line plot suggested by the hint in Exercise 1.2 already illustrates an analysis beyond scatterplot. It uses a crude interpolation.

1.3 Case Study: Distribution Diagnostics

We need specific strategies to detect the presence or the violation of structures. We use random numbers to illustrate how these strategies can look. Here we concentrate on the distribution properties. Let us assume that the sequence consists of independent random numbers with a common distribution. How do we check whether this distribution is a uniform distribution? We will ignore a re-scaling to $(\min, \max)$, which might be necessary, but it is a technical detail that does not affect the investigation substantially. So for now we consider the case $\min = 0; \max = 1$.

Realisations of random variables do not allow us to read off the distributions directly. This is our critical problem. We need characterisations of the distribution which allow an empirical inspection. Of course we can view the observations as measures. For n observations $X_1, \ldots, X_n$ we can define the empirical distribution P_n as $P_n = \sum(1/n)\delta_{X_i}$, where δ_{X_i} is the point measure at X_i. Hence,

$$P_n(A) = \#\{i : X_i \in A\}/n.$$

Unfortunately, the empirical distribution P_n of a set of independent observations with common distribution P is in general very different from the original distribution P. Some properties get lost without repair. Infinitesimal properties are among these. So for example P_n is always concentrated on finitely many points; empirical distributions are always discrete, irrespective of the underlying distribution. To analyse distributions based on data, we need functionals that can be determined based on realisations of random variable and that can be compared to the corresponding functionals of theoretical distributions. One strategy is to restrict ourselves to a family of test sets that is treatable empirically.

Example 1.2: Distribution Function

Instead of the distribution P we consider its distribution function $F = F_P$, where

$$F(x) = P(X \leq x).$$

For an empirical distribution P_n of n observations $X_1, \ldots, X_n$, the corresponding empirical distribution function is

$$F_n(x) = \#\{i : X_i \leq x\}/n.$$

Example 1.3: Histogram

We select disjoint test sets $A_j, j = 1, \ldots, J$, covering the range of X. For example, for the uniform distribution on $(0, 1)$ we can choose the intervals

$$A_j = \Big(\frac{j-1}{J}, \frac{j}{J}\Big]$$

as test sets.

Instead of the distribution P we consider the vector $\big(P(A_j)\big)_{j=1,\ldots,J}$. Its graphical representation is called a ***histogram***. The empirical version is the vector $\big(P_n(A_j)\big)_{j=1,\ldots,J}$.

We will discuss this example in some detail. Some general lessons can be drawn from this example. We will take several passes, moving from a naive approach to an elaborate statistical approach.

We take the opportunity here to point out that the histogram depends critically on the choice of the test sets. In particular, if discretisations in the data meet with an unfortunate choice of the test sets, the results may be misleading. As an alternative to the histogram, we can choose to smooth the data.

Example 1.4: Smoothing

We replace each data point by a (local) distribution, that is, we blur the data points somewhat. We can do this using weight functions. These weight functions are called ***kernels*** and denoted by K. We require that the integral over a kernel exists, and conventionally the kernel is normalised so that $\int K(x)dx = 1$. Some commonly used kernels are listed in Table 1.9 and shown in Figure 1.1. For kernels with compact support, the support is chosen to be the interval $[-1, 1]$. (The R convention is to standardise the kernel, so that the standard deviation is 1.)

By shift and scaling each kernel gives rise to the family

$$\frac{1}{h}K(\frac{x - x_0}{h}).$$

For kernels, the scale factor h is called the kernel ***bandwidth***. The kernel scaled by h, is denoted by K_h:

$$K_h(x) = \frac{1}{h}K(\frac{x}{h}).$$

The function

$$x \mapsto \frac{1}{n} \sum_i K_h(x - X_i)$$

results in a smoothed display that can replace (or enhance) the histogram.

For more information, look for the keywords ***smoothing*** or ***kernel density estimation***.

Kernel	$K(x)$
uniform	$1/2$
triangular	$1 - \|x\|$
Epanechnikov (quadratic)	$3/4(1 - x^2)$
biweight	$15/16(1 - x^2)^2$
triweight	$35/32(1 - x^2)^3$
Gauss	$(2)^{-1/2}\ exp(-x^2/2)$

Table 1.9 *Some commonly used kernels. See Figure 1.1.*

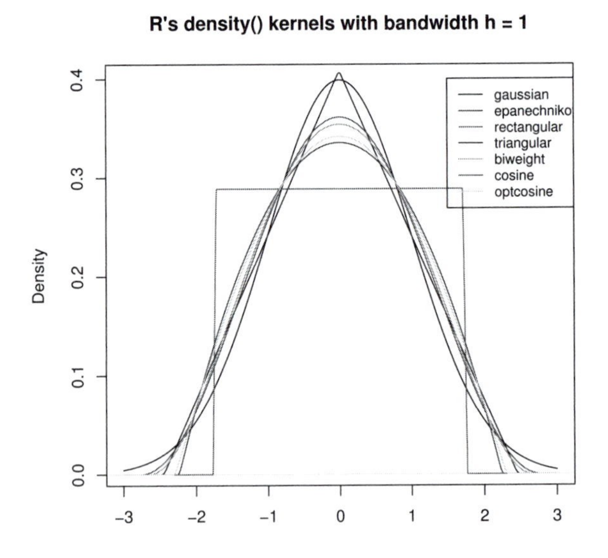

Figure 1.1 *Kernels in* R. *See Example 1.4 on page 11 and Table 1.9. See Colour Figure 1.*

1.3.1 First Pass for Example 1.2: Distribution Functions

To test whether a random sequence has a distribution with distribution function F, we compare F with the empirical distribution function F_n. As a special case, we start with a uniform distribution on $[0, 1]$. In this special case we have $F(x) = F_{unif}(x) = x$ for $0 \leq x \leq 1$. In general, we have to compute the functions F_n and F. A first look tells us F_n is a piece-wise constant function with jumps at the observations. So we get a complete image of F_n if we evaluate F_n at the observations $X_i, i = 1, \ldots, n$. If $X_{(i)}$ denotes the ith order statistic, we have $F_n(X_{(i)}) = i/n$ if $X_{(1)} < \ldots < X_{(n)}$. We have to compare $F_n(X_{(i)})$ with the theoretical value $F(X_{(i)}) = X_{(i)}$. An R implementation, written with temporary variables, is

```
n <- 100
x <- runif(n)
xsort <- sort(x)
i <- (1:n)
y <- i/n
plot(xsort, y)
```

A line with the theoretical values can be added with

```
abline(0, 1)
```

Example 1.5 shows a more compact implementation using the function `length()`.

***Example* 1.5: Empirical Distribution Function**

Input

```
x <- runif(100)
plot(sort(x), 1:length(x)/length(x))
abline(0, 1)
```

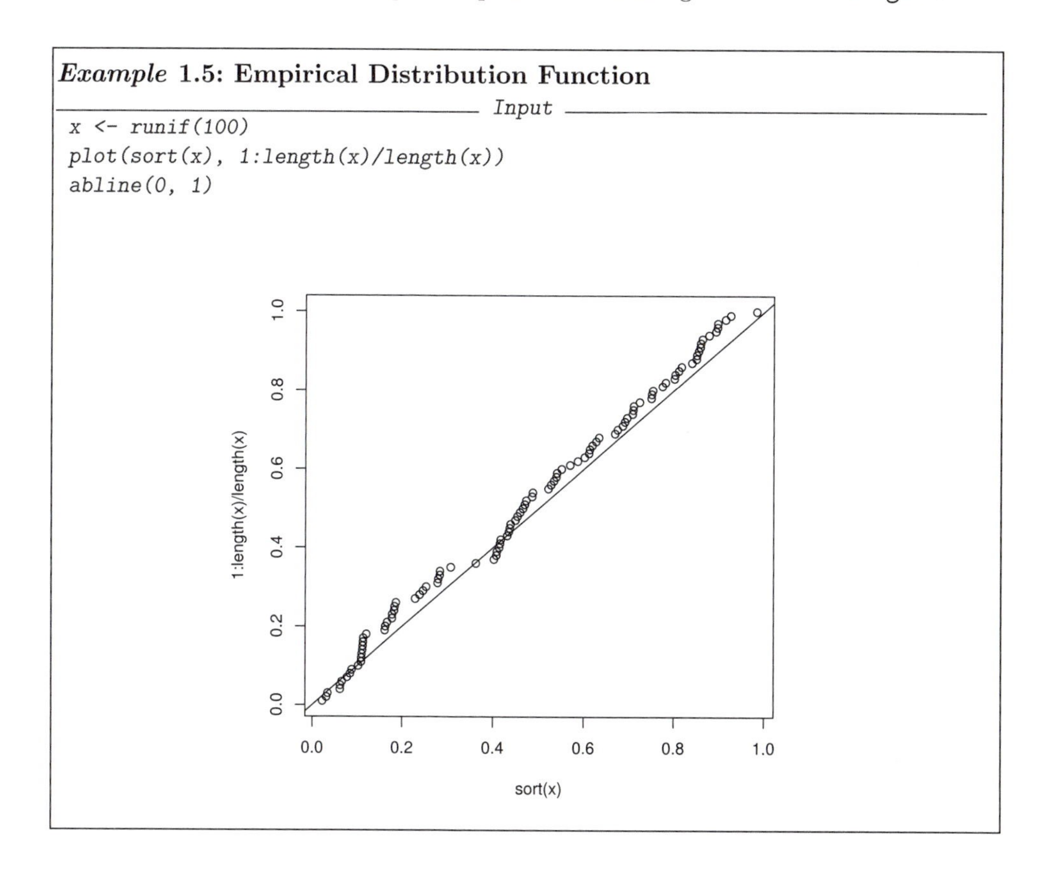

R *Functions*	
`sort()`	sorts a vector. *Example:* `sort(runif(100))`
`length()`	length of a vector. *Example:* `length(x)`
`abline()`	adds a line to a plot. *Example:* `abline(a = 0, b = 2)`

This is still a crude implementation. It does work in general, but may fail in some cases. A more refined implementation would take care of boundary cases or special cases, such as data sets with ties. This sort of detail is what in general is the difference between a first draft and a mature implementation. For now, this implementation serves its purpose.

If we have ties, that is, if we have multiple observations giving the same value, additional precautions have to be taken. The other limitation is that this implementation works for the uniform distribution. In that case, the theoretical distribution function is a diagonal on $[0, 1]$, and this is easily compared with the empirical distribution function. For other distributions, we will modify the idea later on. What we still have to solve in general is how do we judge whether a deviation of an empirical distribution function and a theoretical distribution function is significant?

Exercise 1.3	**Tie Handling**
	What is the result if the data set has ties, that is, multiple entries with the same value? As an example, look at a data set generated by `x <- runif(n); x <- c(x,x)` What should the correct plot look like?
	Try to modify the implementation so that it behaves correctly in the presence of ties. If you cannot implement it with the programming elements introduced so far, give a list of the programming elements you would need to complete this task.

By default, the `plot()` function adds some annotation. To get a graphic speaking for itself we modify the default annotation to show more specific information. We can replace the default arguments of `plot()`. The argument `main` controls the main title (default: empty). We can add a title, for example,

```
plot(sort(x), (1:length(x))/length(x),
      main = "Empirical distribution function\n (X uniform)").
```

Labels for the axis can be controlled using the arguments `xlab` and `ylab`. More information about this and additional arguments can be obtained using `help(plot)`. But the help information only provides a forward reference in these cases. The actual information is given by `help(title)`.

The vertical axis still provides a challenge. Using `ylab = "Fn(x)"` would give a label $Fn(x)$. The usual notation, however, puts the sample size as a subscript, as in $F_n(x)$. A hidden property of the labeling functions helps: if a character string is passed as an argument, it is displayed "as is". If an R expression is passed as an argument, R tries to display it following the usual mathematical conventions. Details are documented in `help(plotmath)` and examples can be seen using `demo(plotmath)`. To convert a character string to an (unevaluated) R expression, use the function `expression()`.

Example 1.6: Empirical Distribution Function (Annotated Plot)

Input

```
x <- runif(100)
plot(sort(x), (1:length(x))/length(x),
    xlab = "x", ylab = expression(F[n]),
    main = "Empirical distribution function\n (X uniform)"
)
abline(0, 1)
```

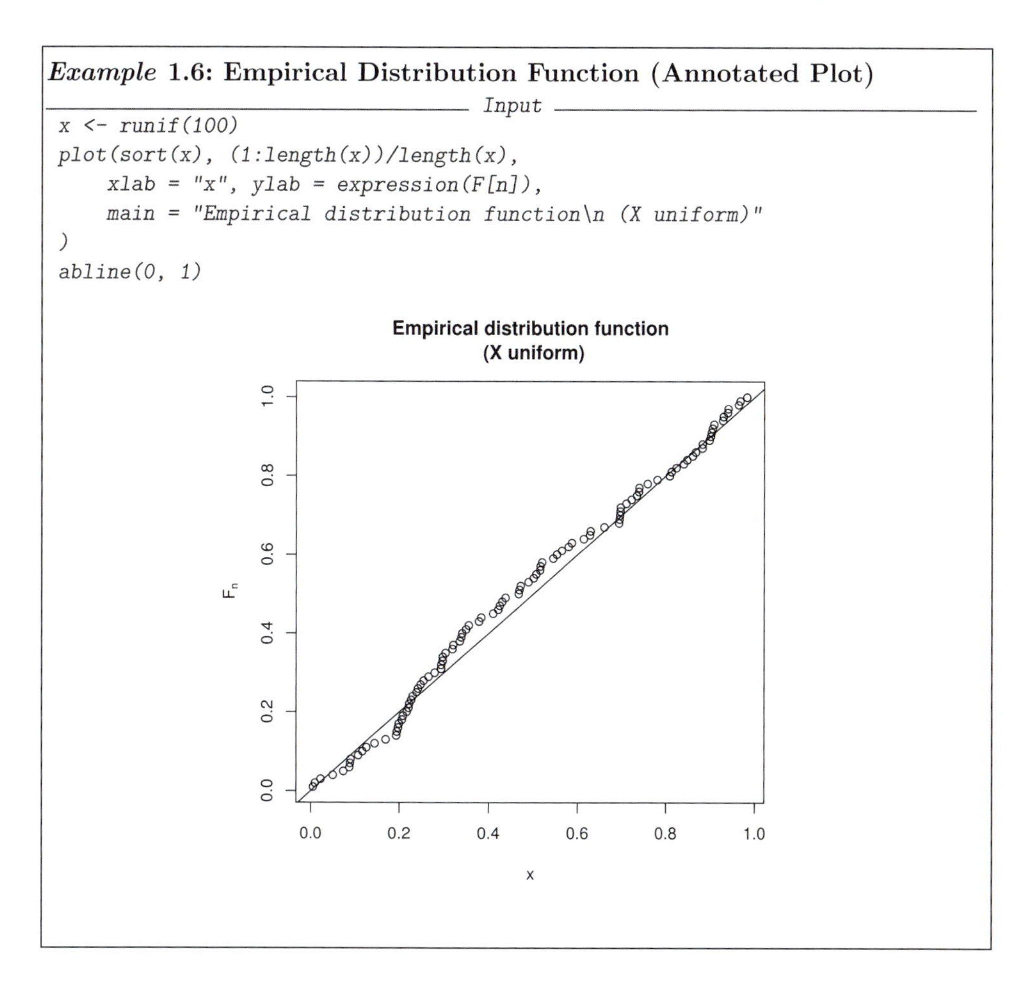

This example serves only as an introduction here. It is not necessary to program distribution functions and their plots from scratch. R already provides a class `ecdf` for empirical distribution functions. If the function `plot()` is applied to an object of class ecdf, the "generic" function *plot()* dispatches internally to the special function *plot.ecdf*, and this draws the distribution function in its specific way. We can reduce our example by just calling *plot(ecdf(runif(100)))*.

Exercise 1.4	
	Extend the function call *plot(ecdf(runif(10)))* using additional arguments so that the result has the following form:

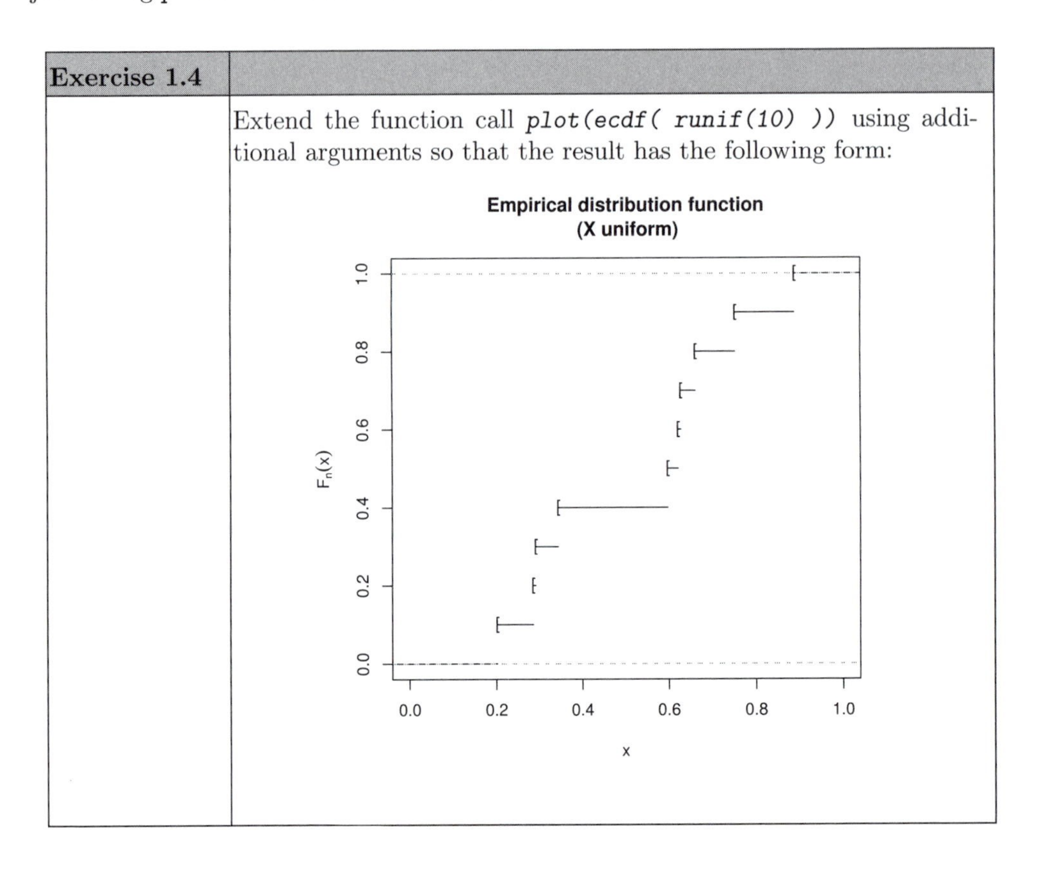

Exercise 1.5	**Discrimination and Sample Size**
	Using norm() you can generate random variables with a Gaussian distribution. Can you tell a Gaussian distribution from a uniform distribution using a serial plot?
	Use the empirical distribution function.
	Can you tell a Gaussian distribution from a uniform distribution using a plot of the empirical distribution function?
	Can you tell the sinus series from a uniform distribution using a plot of the empirical distribution function? from a uniform distribution?
	What is the sample size in each case needed to recognise the differences reliably?

1.3.2 First Pass for Example 1.3: Histograms

We select test sets A_j, $j = 1, \ldots, J$ to cover the range taken by the values of X. Our strategy to test whether the data may correspond to a random sequence from a distribution P, is to compare the vector of theoretical probabilities $(P(A_j))_{j=1,\ldots,J}$ with the vector of observed relative frequencies $(P_n(A_j))_{j=1,\ldots,J}$. For the special case of the uniform distribution on $(0, 1)$ we can take the intervals

$$A_j = \left(\frac{j-1}{J}, \frac{j}{J}\right]$$

as test sets. Then the vector of theoretical probabilities

$$(P(A_j))_{j=1,\ldots,J} = (1/J, \ldots, 1/J)$$

is to be compared to observed relative frequencies $(\#i : X_i \in A_j)/n,\ j = 1, \ldots, J$. The obvious question in this setting is how to choose the number of cells J. A preliminary implementation: we use a ready-made function to draw histograms. As a side effect, it returns the relative frequencies. The function *rug()* adds the original data to the plot.

Example 1.7: Histogram with Rug

Input

```
x <- runif(100)
hist(x)
rug(x)
```

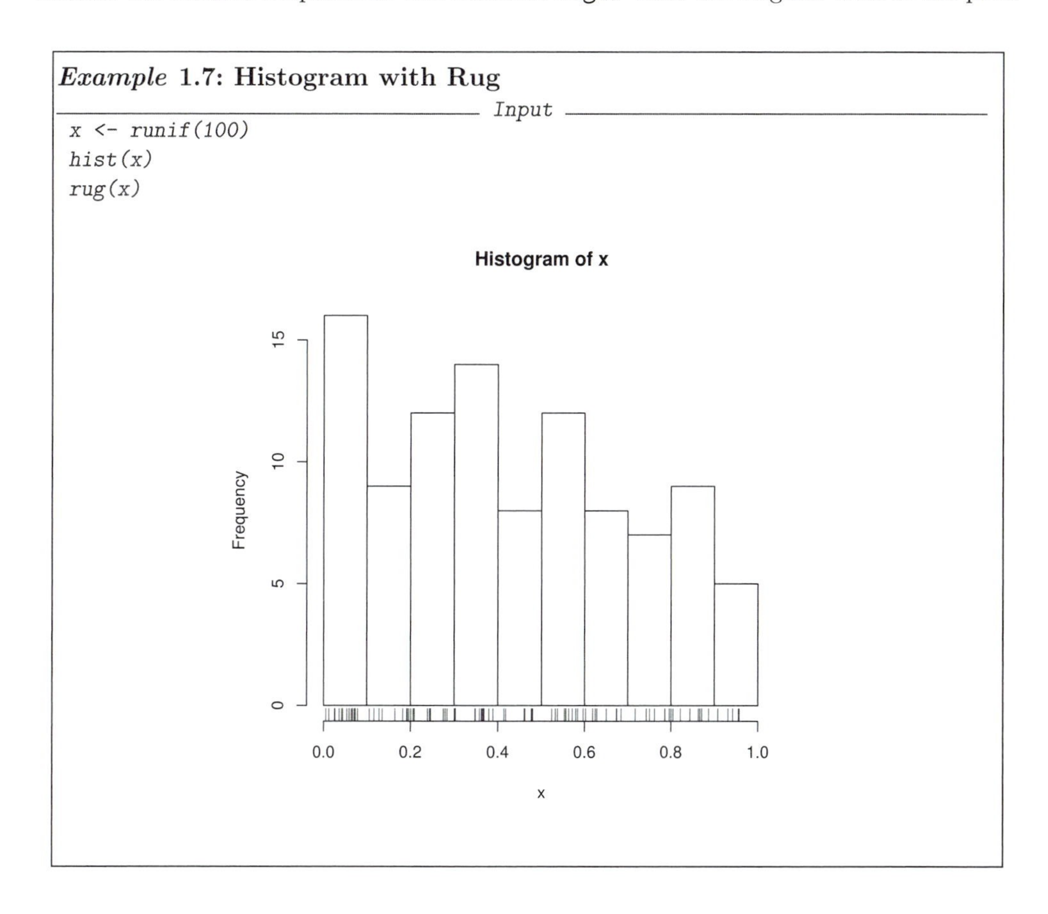

For comparison, we can overlay a kernel density estimation. In contrast to ***hist()***, the function ***density()*** does not draw its result, but prints it. So we have to ask for the graphical output explicitly. To make scales comparable, we use the argument ***probability = TRUE*** to get a histogram using a probability scale.

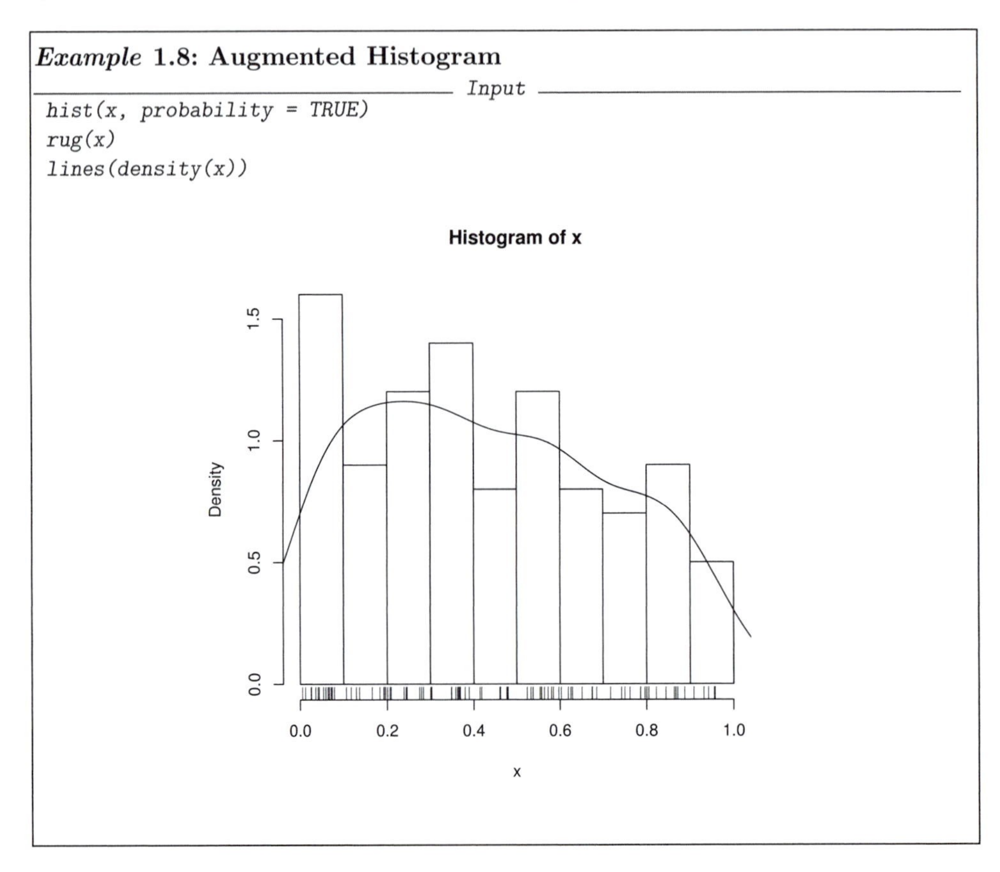

Histograms and kernel density estimators both have their own particularities. Histograms suffer from their discretisation, which may interfere with a possible discretisation of the data and give an aliasing effect. Kernel density estimators do not have this discretisation effect, but they blur the data. This may lead to inappropriate boundary effects.

Let us return to the histogram: If we use an assignment

```
whist <- hist(x),
```

the internal information of the histogram is stored as ***whist*** and can be accessed using

```
whist
```

This gives a result like in the following example:

Example 1.9: Histogram Data Structure

Input
```
x <- runif(100)
whist <- hist(x)
whist
```

Output
```
$breaks
 [1] 0.0 0.1 0.2 0.3 0.4 0.5 0.6 0.7 0.8 0.9 1.0

$counts
 [1] 11 13 13 12  8 14  6  6  9  8

$intensities
 [1] 1.100000 1.300000 1.300000 1.200000 0.800000 1.400000 0.600000 0.600000
 [9] 0.900000 0.800000

$density
 [1] 1.100000 1.300000 1.300000 1.200000 0.800000 1.400000 0.600000 0.600000
 [9] 0.900000 0.800000

$mids
 [1] 0.05 0.15 0.25 0.35 0.45 0.55 0.65 0.75 0.85 0.95

$xname
[1] "x"

$equidist
[1] TRUE

attr(,"class")
[1] "histogram"
```

`Counts` gives the cell counts. So these are exactly the numbers we are looking for. The information stored in `whist` consists essentially of five components. Each of these components is a vector. The components have names and can be accessed using these names. So, for example,

```
whist$counts
```

gives the vector of the cell counts.

Data Structures	
vectors	Components of a vector are accessed by their index. All elements of the vector have the same type. *Examples:* `x` `x[10]`
lists	Lists are complex data structures. The components of a list have names. They can be accessed using these names. Components of a list may be of different types. *Examples:* `whist` `whist$counts`

Additional complex data structures are discussed in Appendix A.8 (page A-202).

The choice of histogram cells is data dependent. There are various conventions and recommendations how to choose cells. The details are controlled by arguments for `hist()`. To select test sets of our choice, we have to inspect the calling structure for `hist()`.

Exercise 1.6	
	Use `runif(100)` to draw random numbers and generate histograms with 5, 10, 20, 50 cells of equal size. Use repeated samples. Do the histogram plots correspond to what you expect from independent uniform random variates? Try to note your observations in detail. Repeat the experiment with two cells $(0, 0.5], (0.5, 1)$. `hist(runif(100), breaks = c(0, 0.5, 1))` Repeat the experiment with random numbers generated by `rnorm(100)` and compare the results from `runif(100)` and `rnorm(100)`.

The plot produced in Example 1.8 tells only half of the story. As far as the histogram is concerned, the histogram bins can be roughly be guessed from the plot. For the kernel density estimation, the choice of the kernel and the chosen bandwidth are critical, and neither can be read from the resulting display. So the plot has to be enhanced to give this essential information in order to be self-contained.

If you compare `hist` and `density`, you will note that R is a developing system. Both functions serve related purposes, but provide different approaches. The obvious difference is that `hist` yields an immediate graphical output while `density` requires an explicit plot request. More subtle differences appear when you try to generate a reproducible analysis. `hist` gives you the details of the histogram used, but does not provide any information on the rules upon which this is based. You have to resort to the history of your commands, and the help information on `hist` (or better yet, the source code of `hist` to be on the safe side) to retrieve this information. `density` returns information on the call you used as part of the result structure, but the kernel used is not recorded in this information. Again, you have to resort to the documentation or the source code.

Exercise 1.7	
	Modify Example 1.8 (page 18) to include the kernel name and the bandwidth used in the kernel density estimation. You have to store the result from `density()` and access its components in analogy to Example 1.9 (page 19).

Barcharts

Note: if the data are not quantitative, but categorical factors (marked by category labels like "excellent, good, satisfactory, ...", or marked by category number, like "1, 2, 3,"), a bar plot may be more appropriate than a histogram. Bar plots are provided by a function `barplot()`. To use `barplot()`, the raw data must be broken down to frequencies for each level of the variable. This can be done using `table()`.

***Example* 1.10: Barchart**

Input

```
grades <- c(2, 1, 3, 4, 2, 2, 3, 5, 1, 3, 4, 3, 6)
barplot(table(grades), xlab = 'Grade', ylab = 'Count',
   ylim = c(0, max(table(grades))),
   main = 'Grading')
```

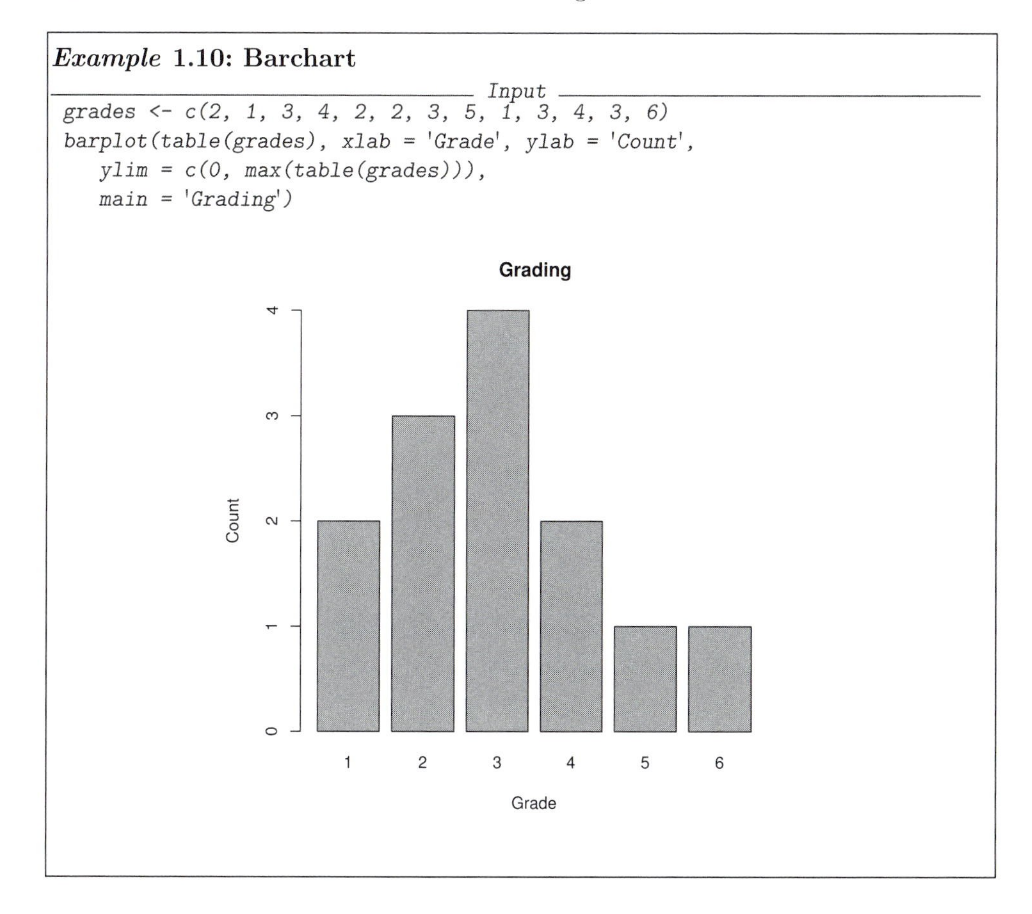

1.3.3 Statistics of Distribution Functions; Kolmogorov-Smirnov Tests

We now move from a naive approach to a statistical point of view. For independent identically distributed variables $(X_1, \ldots, X_n)$ with distribution function F, in a naive approach we have assumed that $i/n = F_n(X_{(i)}) \approx F(X_{(i)})$. We wanted to use this relation to test our distribution assumption. In particular, for a uniform distribution on $(0, 1)$ this relation reads $i/n \approx X_{(i)} = F(X_{(i)})$.

From a statistical point of view, each $X_{(i)}$ is a random variable. Hence $F(X_{(i)})$ is also a random variable. Moreover, if F is continuous, $F(X_{(i)})$ are independent random variables with a uniform distribution on $[0, 1]$. We can analyse the distribution of these random variables.

Theorem 1.1 *If $(X_1, \ldots, X_n)$ are independent random variables with common continuous distribution function F, the $F(X_{(i)})$ has a beta distribution $\beta(i, n-i+1)$.*

Proof. $\rightarrow$ probability theory. *Hint:* Use

$$X_{(i)} \leq x_\alpha \Leftrightarrow (\#j : X_j \leq x_\alpha) \geq i.$$

For continuous distributions, $(\#j : X_j \leq x_\alpha)$ has a binomial distribution with parameters (n, α). $\square$

Corollary 1.2 $E(F(X_{(i)})) = i/(n+1)$.

Exercise 1.8	
	Using `help(rbeta)` you get information about the functions available to work with beta distributions. Generate plots for the densities of the beta distribution for $n = 16, 32, 64, 128$ and $i = n/4, n/2, 3n/4$. Use the function `curve()` to generate the plots. For more information, see `help(curve)`.

So on average, for independent random variables with uniform distribution on $(0, 1)$ we cannot expect that $X_{(i)} \approx i/n$, but on average we get $i/(n+1)$, the expected value of the beta distribution. Hence the reference line should be drawn with `abline(a = 0, b = n/n+1)`.

Exercise 1.9	
	Draw the distribution function with the corrected reference line.
*	We use the graphical display for a single sample, not for a run of samples. Is the expected value of $X_{(i)}$ an adequate reference? Are there alternatives that can serve as references? If you see alternatives, give an implementation.

Monte Carlo Confidence Bands

Simulation can also help us to get an impression of the typical fluctuation. We use random numbers to generate a small number of samples, and compare our sample in question with these simulations. For comparison, we generate envelopes of these simulations and check whether our sample lies within the area delimited by the envelopes. If x is the sample vector to be analysed, with length n, we use this programming idea:

***Example* 1.11: Monte Carlo Confidence Bands**

Input

```
x <- (sin(1:100)+1)/2                         # demo example only
y <- (1:length(x))/length(x)
plot(sort(x), y)
nrsamples <- 19                               # no of simulations
samples <- matrix(data = runif(length(x)* nrsamples),
    nrow = length(x), ncol = nrsamples)
samples <- apply(samples, 2, sort)
envelope <- t(apply(samples, 1, range))
lines(envelope[, 1], y, col = "red")
lines(envelope[, 2], y, col = "red")
```

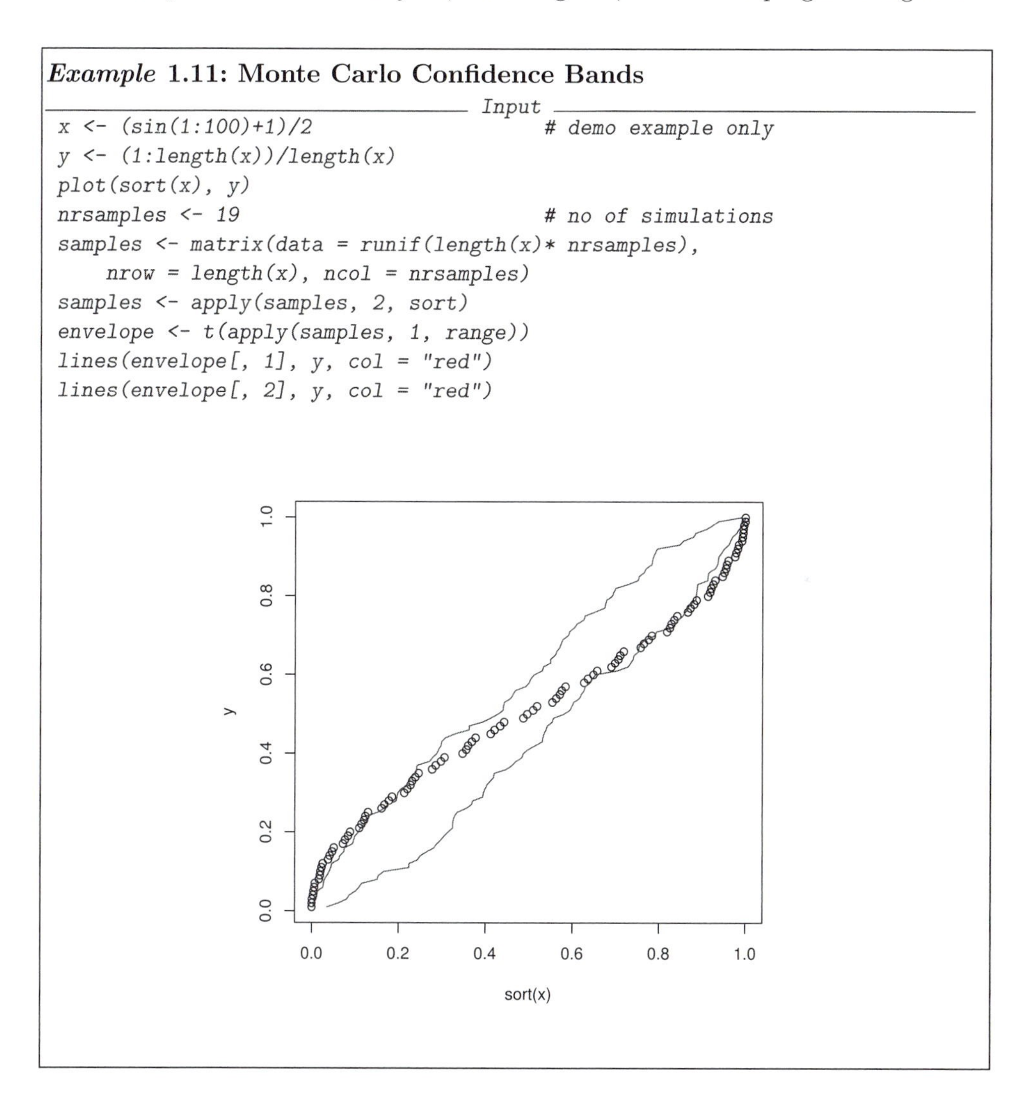

This example is taken from [52], which is a rich source of ideas for R programming.

A typical R programming strategy is illustrated here. R is an interpreted vector-oriented language. Single-step interpretation is time intensive. Hence operations with a few complex steps can be more efficient than operations using several elementary steps.

- Operations at a vector level are more effective than chains of single elementary operations.
- Iterations and loops should be avoided and replaced by structured vector operations. See [25].

Exercise 1.10	
	Make use of the `help()`-function and comment on Example 1.11 step by step. Take special note of the new functions that are introduced here.

R *Iterators*	
`apply()`	applies a function to the rows or columns of a matrix. *Example:* `samples <- apply(samples, 2, sort)` sorts by column.
`outer()`	generates a matrix of all pair-wise combinations of two vectors, and applies a function to all pairs.

See Appendix A.9 (page A-204) for other iterators provided with R.

For the simulation bands, the hypothesis is that the given sample and the simulations come from the same model. If the curve representing our sample exceeds the limits from our simulation, this indicates that this hypothesis is violated. The approach that is sketched here is called a ***Monte Carlo test***. The idea behind this is very general and can be applied in many situations. For the theoretical background, see [39], Section 7.1.

Exercise 1.11	
*	Why 19? *Hint:* Try to take an abstract simplified view of the problem first: let T be a measurable function and $X_0, X_1, \ldots, X_{nrsamples}$ independent samples with a common distribution function. What is $P(T(X_0) > T(X_i))$ for all $i > 0$? In a second step, give an abstract formulation for the example above. Then consider the special case `nrsamples = 19`.

Exercise 1.12	**Monte Carlo Coverage**
*	Estimate the coverage probability of the Monte Carlo band by first generating a band as above. (How can you draw the band without first making a plot for a special sample?) Next, generate *sim* simulation samples of uniform random numbers of sample size 100. Count how many simulations give a sample within the band. You have to make your choice of the number *sim* of simulations (100? 1000? 999?) for this step. Use this information to estimate the coverage probability. *Hint:* `any()` can be used to evaluate a comparison for a full vector.

As usual, we want to rework the output so that the resulting plot contains enough information. For the legend, we can proceed as in Section 1.3.1 (page 13). The simulation count `nrsamples` requires some thought. If only a fixed count (e.g., 19) is considered, we can use the constant in the legend as usual. If the program fragment must be more versatile, we would need the specific simulation count for each instance. Doing this by hand is prone to error, and this possible source of errors can be avoided. The function `bquote()` allows us to evaluate a variable, or to calculate it for a specific context. To allow us to evaluate the current number of simulations, we rearrange the statements so that the number of observations is fixed before `plot()` is called.

We start by providing some test data:

Input

```
x <- (sin(1:100)+1)/2                    # demo example only
y <- (1:length(x))/length(x)
nrsamples <- 19                         # nr of simulations
```

***Example* 1.12: Monte Carlo Confidence Bands (Augmented)**

Input

```
plot(sort(x), y,
    main = paste("Monte Carlo Band: ",
        bquote( .(nrsamples)), " Monte Carlo Samples"),
        xlab = 'x', ylab = expression(F[n]))
samples <- matrix(data = runif(length(x) * nrsamples),
    nrow = length(x), ncol = nrsamples)
samples <- apply(samples, 2, sort)
envelope <- t(apply(samples, 1, range))
lines(envelope[, 1], y, col = "red")
lines(envelope[, 2], y, col = "red")
```

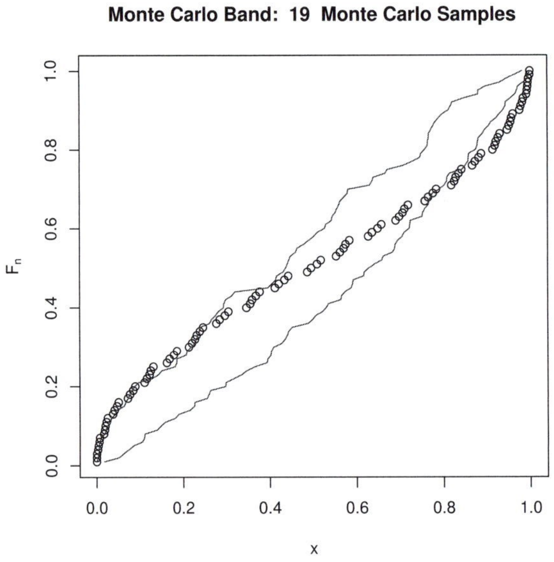

For each simulation new Monte Carlo samples are drawn. So for each invocation we get different Monte Carlo bands, and the bands here are different from those in the previous example.

For practical situations it may be necessary to reduce distribution diagnostics to a simple decision problem. This may allow us to produce tables or control charts to decide whether a distribution fits to some hypothesis, or to characterise the deviation from a given model. If we want to use tables or simple numbers, additional restrictions are needed: we need to reduce the information contained in the functions (F_n, F) to simple numbers. One digest is, for example, the statistic

$$\sup_x |F_n - F|(x).$$

If we want to use this statistic as a criterion, we once again are faced with the task of analysing its distribution.

Theorem 1.3 *(Kolmogorov, Smirnov) For a continuous distribution function F, the distribution of*

$$\sup_x |F_n - F|(x)$$

is independent of F (in general, it will depend on n).

Proof. $\rightarrow$ probability theory. For example [46] or [11], Lemma 3.3.8 or [9]. $\square$

Theorem 1.4 *(Kolmogorov): For a continuous distribution function F and $n \rightarrow \infty$ the statistic*

$$\sqrt{n} \sup |F_n - F|$$

has asymptotically the distribution function

$$F_{Kolmogorov-Smirnov}(y) = \sum_{m \in \mathbb{Z}} (-1)^m e^{-2m^2y^2} \qquad \text{for } y > 0.$$

Proof. $\rightarrow$ probability theory. For example [46] or [11], Formula (3.3.11) or [9]. $\square$

Practically speaking, this means that for a continuous distribution function we can use the following decision strategy: we decide that observations $(X_1, \ldots, X_n)$ do not follow the hypothesis of independent identically distributed observations with the distribution function F, if $\sup |F_n - F|$ is too large:

$$\sup |F_n - F| > F_{crit}/\sqrt{n},$$

where F_{crit} is taken from the distribution function of the Kolmogorov-Smirnov statistic for sample size n. This distribution function does not depend on F. In particular, if we choose the upper α quantile $F_{crit} = F_{Kolmogorov-Smirnov,1-\alpha}$, we know that if the hypothesis is valid, this critical value is reached or exceeded at most with probability α. So we can control the error probability of a false rejection of the hypothesis.

Asymptotically, for large n, we can use the Kolmogorov-Smirnov approximation instead of the distribution function. If the model distribution F is not continuous, additional considerations are necessary.

Asymptotics is convenient from a mathematical point of view (because you can use limit theorems), and this is what you find in most textbooks. But if it comes to practical applications, where $n = 10$, or $n = 100$, or even $n = 1000$ or even a larger but always a finite number, asymptotics always must be viewed with caution. What is the finite sample behaviour? In the case of the Kolmogorov-Smirnov statistic, fortunately, we have not only asymptotics, but also bounds that apply to finite sample size and become effective (non-trivial) for very small numbers.

Theorem 1.5 *For all integer n and any positive λ, we have*

$$P(\sqrt{n}\sup|F_n - F| > \lambda) \leq 2e^{-2\lambda^2}.$$

Proof. [27], Corollary 2 □

This inequality is valid even if F is not continuous.

Exercise 1.13	Finite Sample Bounds
	Use the inequality given in Theorem 1.5 to calculate bounds for $\sqrt{n}\sup|F_n - F|$.
	Add finite sample bands to the empirical distribution function.

We want to concentrate on programming issues and not the details of the Kolmogorov-Smirnov tests. With elementary tools, we can calculate the test statistic $\sup_x |F_n - F|(x)$ for the uniform distribution. By monotonicity,

$$\sup_x |F_n - F|(x) = \max_{X(i)} |F_n - F|(X(i))$$

and for the uniform distribution

$$\max_{X(i)} |F_n - F|(X(i)) = \max_i |i/n - X(i)|.$$

In R notation, the expression

```
max( abs((1: length(x)) / length(x)) - sort(x)) )
```

is the statistic we are looking for, based on a data vector `x`.

Like many commonly used statistics, this statistic and the corresponding distribution function are already implemented in R.[2]

Exercise 1.14	

	Using `help(ks.test)` you get information on how to invoke the function `ks.test`.
	Which results do you expect if you test the following vectors for a uniform distribution?
	`1:100`
	`runif(100)`
	(cont.)→

[2] Different implementations can use other calling structures. The Kolmogorov-Smirnov test is available as `ks.test()`. The distribution function of the test statistic is hidden in the internals of R. Before R version 2.x, however, `ks.test()` was not included in the base of R, but was contained in special libraries that had to be loaded explicitly. In R1.x, for example, the package of classical tests needed to be loaded by `library(ctest)`.

Exercise 1.14	(cont.)

	`sin(1:100)`
	`rnorm(100)`
	Perform these tests and discuss the results. For the test, scale the values so that they fall into the interval $[0,1]$, or use a uniform distribution on an interval that is adapted to the data.

1.3.4 Statistics of Histograms and Related Plots; χ^2-Tests

As we did with the distribution functions, we now move towards a statistical analysis of histograms. For simplicity we assume that we have chosen $A_j, j = 1, \ldots, J$ covering the range of X. The vector of observations $(X_1, \ldots, X_n)$ translates into bin counts N_j

$$N_j = (\#i : X_i \in A_j).$$

If $(X_i)_{i=1,\ldots,n}$ are independent with identical distribution P, $(N_j)_{j=1,\ldots,J}$ is a random vector following a multinomial distribution with parameters $n, (p_j)_{j=1,\ldots,J}$ where $p_j = P(A_j)$. For the special case $J = 2$ we have the binomial distribution. Since we are free in our choice of the sets A_j, we can cover a spectrum of special cases, like

Median test for symmetry:

$$A_1 = \{x < x_{0.5}\} \qquad A_2 = \{x \geq x_{0.5}\}$$

Midrange test for concentration:

$$A_1 = \{x_{0.25} \leq x < x_{0.75}\} \qquad A_2 = \{x < x_{0.25} \text{ or } x \geq x_{0.75}\}.$$

For the general case, however, we must compare the empirical vector of bin counts N_j with the multinomial distribution, which is not pleasant to calculate. We resort to using approximations. The following approximation goes back to Pearson:

Lemma 1.6 *(Pearson): For $(p_j)_{j=1,\ldots,J}$, $p_j > 0$ in the limit $n \to \infty$ the following approximation holds:*

$$\begin{aligned} P_{mult}(N_1, \ldots, N_j; n, p_1, \ldots, p_j;) \approx \\ (2\pi n)^{-\frac{1}{2}} \Big(\prod_{j=1,\ldots,J} p_j \Big)^{-\frac{1}{2}} \cdot \\ \exp\Big(-\frac{1}{2} \sum_{j=1,\ldots,J} \frac{(N_j - np_j)^2}{np_j} \\ -\frac{1}{2} \sum_{j=1,\ldots,J} \frac{N_j - np_j}{np_j} \\ +\frac{1}{6} \sum_{j=1,\ldots,J} \frac{(N_j - np_j)^3}{(np_j)^2} + \ldots \Big). \end{aligned}$$

Proof. → probability theory. For example [20], p. 285. □

The first term in the exponent of e is controlled by $\chi^2 := \sum_{j=1\ldots J}(N_j - np_j)^2/np_j$. This term is called the χ^2 statistic. At least asymptotically for $n \to \infty$ large values of χ^2 give small probabilities. This motivates us to use the χ^2 statistic approximatively as a measure for goodness of fit. The name comes from the distribution asymptotics:

Theorem 1.7 *(Pearson): for $(p_j)_{j=1,\ldots,J}, p_j > 0$ in the limit, as $n \to \infty$, the statistic*

$$\chi^2 := \sum_{j=1,\ldots,J} \frac{(N_j - np_j)^2}{np_j}$$

has a χ^2 distribution with $J-1$ degrees of freedom.

As above, to get a formal decision rule we can fix a critical value χ^2_{crit} and reject the hypothesis that the observations $(X_1, \ldots, X_n)$ come from independent identically distributed uniform random variables if the χ^2 statistic exceeds this value. If we choose the upper α quantile of the χ^2 distribution as a critical value, we know that, if the hypothesis holds, the value χ^2_{crit} or a higher value is reached at most with probability α. So again, at least asymptotically, we can control the error probability of a false rejection of the hypothesis.

The χ^2 tests are a basic part of R and provided as the function `chisq.test()`. As always, the asymptotics used for the "standard" χ^2 tests has to be viewed with caution. Even theoretically, the conditions are more delicate here. It is not sufficient that the sample size is large, but the asymptotics requires that the sample size in each histogram cell must be large. As a second possibility, besides using the asymptotic distribution, `chisq.test()` provides the possibility of calculating compute p-values by Monte Carlo simulation (see `help(chisq.test)`).

The χ^2 tests are designed to be used for more general "contingency tables". We just need a special case. In our case, the table is a (one-dimensional) vector of counts for the chosen histogram cells. (See also the R implementation for more general variants in `loglin()`.)

Exercise 1.15	
	Use `help(chisq.test)` to see the calling structure for χ^2 tests. Apply it to test the hypothesis $(p_j = 1/J)$, $J = 5$ on the following vectors of bin counts: $$(3\ 3\ 3\ 3\ 3) \qquad (1\ 2\ 5\ 3\ 3) \qquad (0\ 0\ 9\ 0\ 6).$$

Exercise 1.16	

	Which results do you expect if you use a χ^2 test to check the following vectors for a uniform distribution?
	`1:100`
	`runif(100)`
	`sin(1:100)`
	`rnorm(100)`
	Perform these tests and discuss the results.
	Hint: The function `chisq.test()` expects a frequency table as input. The function `table()` can be used to generate a frequency table directly (see `help(chisq.test)`). But you can also use the function `hist()`, which gives `counts` as one component of its result.

At first, the approximation for the χ^2 statistic is only valid if fixed cells are chosen, independent of the information provided by the sample. Practical histogram algorithms, however, derive the number of cells and cell bounds based on the sample. Often this implies an estimation of parameters of the distribution. Under some conditions, however, the χ^2 asymptotic still holds, as, for example, illustrated by the following theorem from [36] Section 6b.2:

Theorem 1.8 *(i) Let the cell probabilities be the specified functions $\pi_1(\boldsymbol{\theta})$, ..., $\pi_k(\boldsymbol{\theta})$ involving q unknown parameters $(\theta_1, \ldots, \theta_q) = \boldsymbol{\theta}'$. Further, let*

(a) $\widehat{\boldsymbol{\theta}}$ be an efficient estimator of $\boldsymbol{\theta}$ in the sense of [36] (5c.2.6),

(b) each $\pi_i(\boldsymbol{\theta})$ admit continuous partial derivatives of the first order (only) with respect to θ_j, $j = 1, \ldots, q$ or each $\pi_i(\boldsymbol{\theta})$ be a totally differentiable function of $\theta_1, \ldots, \theta_q$, and

(c) the matrix $M = (\pi_r^{-1/2} \partial\pi_r / \partial\theta_s)$ of order $(k \times q)$ computed at the true values of $\boldsymbol{\theta}$ be of rank q. Then the asymptotic distribution of

$$\chi^2 = \sum \frac{(N_i - n\widehat{\pi}_i)^2}{n\widehat{\pi}_i} \tag{1.1}$$

is $\chi^2(k-1-q)$, where $\widehat{\pi}_i = \pi_i(\widehat{\theta})$.

Proof. See [36], Section 6b.2. □

Exercise 1.17	
*	Sketch comparable test environments for fixed and adaptive choice of histogram cells. (cont.)→

Exercise 1.17	(cont.)
	For fixed and for adaptive choice of histogram cells draw `s = 1000` samples of size 50 from *runif()*. Calculate in both settings the formal χ^2 statistics and plot its distribution functions. Compare the distribution functions.

Different choices of the number of cells and breakpoint positions are possible, depending on the purpose of the histogram. One application is to use the envelope curve of the histogram as a density estimator. The built-in algorithms for breakpoint selection of *hist()* are motivated by this application. As a density estimator, the kernel density estimators (Example 1.4) are direct competitors. Kernel density estimators offer more flexibility (for example, to control smoothness) whereas in general histograms are easier to calculate. More often, histograms are used to detect ***data features***. However, unless the histograms are tailored for special cases, more efficient plots are available for this purpose. We will return to the topic of data features in Section 3.6.

Repeated Samples

So far, we have concentrated on inspecting the distribution of random variables. We can continue this inspection. If $(X_1, \ldots, X_n)$ are independent random numbers with identical uniform distribution, for fixed cells the χ^2 statistics is approximatively distributed as χ^2, and $\kappa := \sqrt{n} \sup |F_n - F|$ has the Kolmogorov-Smirnov distribution.

We can take repeated samples $(X_{1,j}, \ldots, X_{n,j})_{j=1,\ldots,m}$ and from them we can calculate statistics ${\chi^2}_j$ and κ_j. Given independent, identically distributed random variables to start with, the statistics have a χ^2 resp. a Kolmogorov-Smirnov distribution. Taking these repeated samples we can not only inspect the distribution of the single observations, but also the joint distribution of samples, each consisting of n sample elements.

Exercise 1.18	
	For $n = 10, 50, 100$, draw 300 samples using *runif(n)*. For each sample, calculate the χ^2 and the Kolmogorov-Smirnov statistic. You have to choose a χ^2 test. What is your choice? Plot the distribution functions of these statistics and compare them to the theoretical (asymptotic) distributions. Are there any indications against the assumption of independent uniform random numbers? *Hint:* The functions for the χ^2 and Kolmogorov-Smirnov test keep their internal information as a list. To get the names of the list elements, you can create a sample object. For example, use *names(chisq.test(runif(100)))*.

Power

So far in our discussion the uniform distribution was the target or model distribution, our "hypothesis". We have discussed how various methods will behave, if this hypothesis holds. The distribution properties derived from the model assumption can indicate how to find critical values for formal tests. We reject the hypothesis, if the observed test statistics are too extreme. What "too extreme" means is determined by the distributions we derived from the hypothesis. This leads to decision rules like:

reject the hypothesis, if $F_{\chi^2}(\chi^2) \geq 1 - \alpha$

or

reject the hypothesis, if $F_{Kolmogorov-Smirnov}(\kappa) \geq 1 - \alpha$

for chosen (small) values of α.

In the next step, if we have fixed a formal decision rule, we can ask for the power of this rule to reject the hypothesis if it actually does not hold. A more precise analysis of the power of decision rules is a theme in classical lectures in statistics. The possibilities discussed so far allow us to analyse the performance using Monte Carlo strategies.

As a framework for simulations we choose a family of alternatives. The uniform distribution blends into the family of beta distributions with densities

$$p_{a,b}(x) = \frac{\Gamma(a+b)}{\Gamma(a)\Gamma(b)} x^{a-1}(1-x)^{b-1} \quad \text{for } a > 0, b > 0 \text{ and } 0 < x < 1.$$

We choose alternative distributions from this family. From these we draw repeated samples, and apply our decision rule formally to each of these.

We note whether the procedure leads to a rejection of the hypothesis or not. Given a choice of the sample size n and a number of simulations m, and given a choice of a limiting probability α we get the table

$$(a, b) \mapsto \# \text{ simulations leading to a rejection of the hypothesis.}$$

In particular, for the uniform distribution $(a, b) = (1, 1)$ we expect approximately $m \cdot \alpha$ rejections. For distributions other than the uniform hypothesis, a decision procedure is more powerful if the proportion of rejections is higher.

Exercise 1.19	
**	Analyse the power of the Kolmogorov-Smirnov test and the χ^2 tests. Select values for n, m and α, and choose 9 pairs for (a, b). What are your arguments for your choices? Use your chosen parameters to draw samples from `rbeta()`. Apply the Kolmogorov-Smirnov test and a χ^2 test with 10 cells of equal size on $(0, 1)$.
	(cont.)→

Exercise 1.19	(cont.)
	Choose alternative parameters (a, b) so that you can compare the decision rules along the following lines: i) a = b ii) b = 1 iii) a = 1 and run these simulations.
	Choose alternative parameters (a, b) so that you can compare the decision rules over the range $0 < a, b < 5$. Your conclusions? *Hint:* `outer(x, y, fun)` applies a function `fun()` to all pairs of values from x, y and returns the result as a matrix. Using `contour()` you can generate a contour plot. See `demo("graphic")`.

Exercise 1.20	
⋆⋆	Design a test strategy to unmask "pseudo-random numbers". Test this strategy using simple examples i) $x \quad x = 1..100 \mod m$ for convenient m ii) $\sin(x)$ $x = 1..100$ iii) ... Do you tag these sequences as "not random"? Now try to unmask the random number generators provided by R. Can you identify the generated sequences as "not random"?

1.4 Moments and Quantiles

Distribution functions or densities are not easy objects from a mathematical point of view. In general, the space of functions is infinite-dimensional. Finite-dimensional geometrical or optimisation arguments cannot be applied directly. To simplify analysis one often resorts to finite-dimensional descriptions.

Historically, moments have played an important role: probability distributions are viewed as mass distributions, and the moments are defined in analogy to the moments in mechanics. The first moment, corresponding to the center of gravity, is called ***expected value*** in statistics.

Definition 1.9 *If X is a real valued random variable with distribution P, the expected value of X is defined as*

$$E_P(X) := E(X) := \int X dP$$

provided this integral exists.

The second moment and higher moments are usually centred. For the second (central) moment, the ***variance***, we have the following definition:

Definition 1.10 *If X is a real valued random variable with distribution P, its variance is defined as*

$$Var_P(X) := Var(X) := \int (X - E(X))^2 dP$$

provided this integral exists.

The integral does not always give defined values. Hence the moments do not always exist. But if they exist, they give the first information about the distribution. The expected value is often interpreted as the "statistical average" or mean of the distribution. The root of the variance, the ***standard deviation***, is interpreted as the "statistical spread".

The definitions can be applied to empirical distributions as well. This gives the first possibility of estimating the moments of an unknown theoretical distribution from the data. For the expected value, the mean, we have an unbiased estimator:

$$E_P(\ E_{P_n}(X)\ ) = E_P(X),$$

that is, in the statistical average the empirical mean and the mean of the underlying theoretical distribution coincide (provided they are defined).

For the variance the unbiasedness does not hold, but we have

$$\frac{n}{n-1} E_P(\ Var_{P_n}(X)\ ) = Var_P(X),$$

for $n > 1$. The mathematical background is that the expected value is a linear operator. It commutes with linear operators. But the variance is a quadratic operator, and that requires a correction if we want an unbiased estimator. The variance with this bias correction is often called the ***empirical variance*** or ***sample variance***.

For the estimation of the first two moments of a vector of random numbers, R provides two functions: `mean()` estimates the mean and `var()` estimates the variance, using the sample variance. The function `sd()` gives the standard deviation of a vector.

A quite different approach is to reduce the distribution function to a small set of percentage points, or ***quantiles***, such as the median (the 50% point) and the lower and upper quartile (the 25% and 75% point). With this approach, for example the ***interquartile range*** (the difference between lower and upper quartile) can be used as a measure of spread. Since distribution functions can have steps and flat parts, an exact definition of quantiles has to be more elaborate:

Definition 1.11 *$x \in \mathbb{R}$ is a p* **quantile** *of X, if $F_X(x) \geq p$ and $F_X(x') \leq p$ for all $x' < x$.*

Exercise 1.21	
	Generate a sample of random variables with sample size 100 from the distributions with the following densities: $$p(x) = \begin{cases} 0 & x < 0 \\ 1 & 0 \leq x \leq 1 \\ 0 & x > 1 \end{cases}$$ and $$p(x) = \begin{cases} 0 & x \leq 0 \\ 2 & 0 < x \leq 1/4 \\ 0 & 1/4 < x \leq 3/4 \\ 2 & 3/4 < x \leq 1 \\ 0 & x > 1 \end{cases}$$ Estimate the mean, variance and standard deviation in each of these. Repeat the estimation for 1000 samples. Analyse the distribution of estimated mean, variance and standard deviation for repeated samples.

Moments can be calculated using simple arithmetic operations. Their combination (exact or approximate) follows simple laws; however, they are quite sensitive. Even shifting minimal probability mass can lead to a breakdown. For the empirical distribution this means if a proportion of $1 - \varepsilon$ of the observed data follows a model distribution and a proportion of ε comes from some different distribution, the moments can take any value, even for arbitrarily small values of ε. Quantiles are much more robust against a breakdown. At least 50% of the data have to be "outliers" to lead to a breakdown of the median, while changes in just one data point can give arbitrary value to the mean.

With the availability of sufficient computing power, quantiles have gained more importance as descriptors. Calculating quantiles implicitly requires sorting. Hence it is more complex than the calculation of moments. Rules for combination of quantiles are not as simple as those for moments and often need explicit calculations. But with the technical resources now available this is not an essential restriction.

R provides several functions to work with quantiles.

`quantile()` is an elementary function to calculate quantiles. The function `summary()` gives a summary of the distribution information, which includes information about quantiles.

Exercise 1.22	
	Generate a sample of 100 random variables from the distributions of Exercise 1.21. Estimate the median, and the lower and upper quartiles. Repeat the estimation for 1000 samples. Analyse the distribution of the estimated median, lower and upper quartiles from repeated samples.

`boxplot()` gives a graphical representation of a quantile-based summary. The "box-and-whisker plot" that is used here has many variations. So for interpretation of a box-and-whisker plot it is necessary to get detailed information about which convention is used. Usually a box is used to mark the central part of the distribution. In the standard version a line is used to mark the median, and a "box" goes from the median of the upper half to the median of the lower half. This corresponds roughly to the upper and lower quartile. The finer definition ensures that the information is still reliable if the data contain ties, that is, repeated observations with the same values. The "whisker" describes the adjacent area. Outliers are marked.

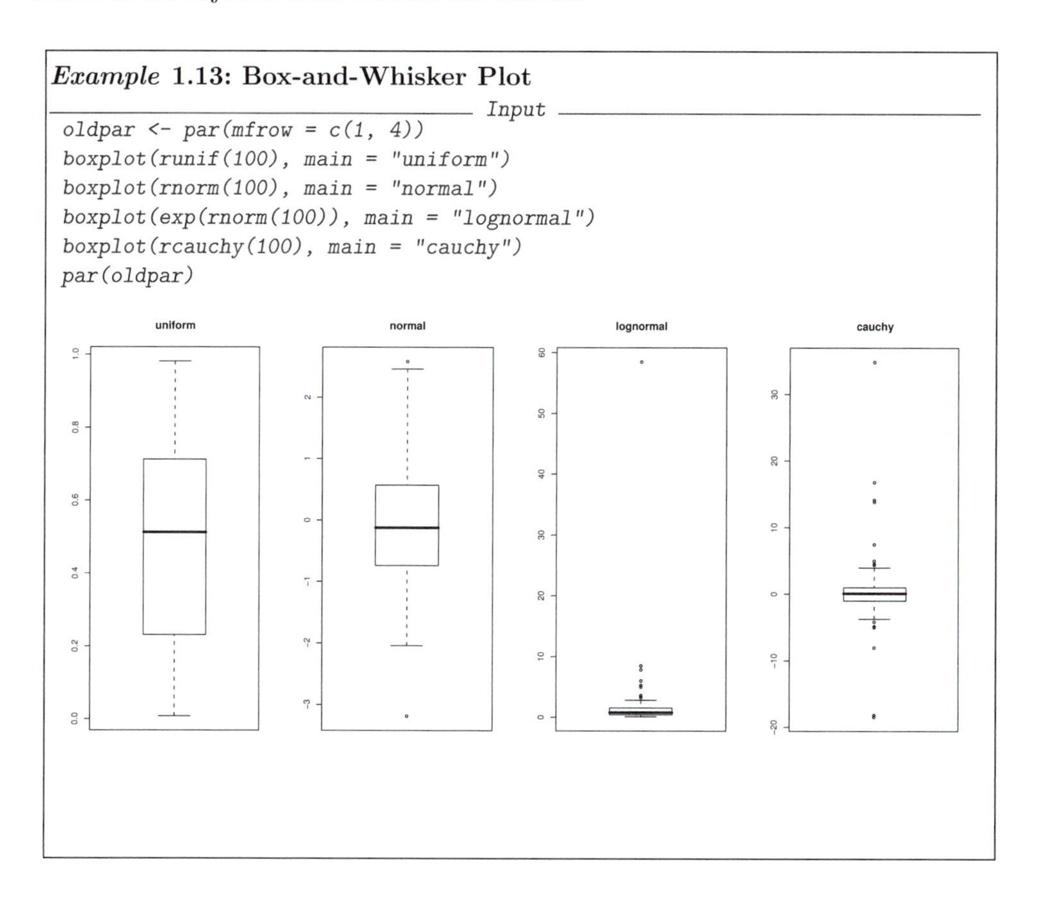

Example **1.13: Box-and-Whisker Plot**

Input

```
oldpar <- par(mfrow = c(1, 4))
boxplot(runif(100), main = "uniform")
boxplot(rnorm(100), main = "normal")
boxplot(exp(rnorm(100)), main = "lognormal")
boxplot(rcauchy(100), main = "cauchy")
par(oldpar)
```

The plot is not made for short-tailed distributions, such as the uniform. To see the difference between the uniform and the normal distribution in these plots requires some training. Both are symmetric, and the only difference that appears in the plot is the different tail behaviour, marked by the length of the whiskers. The lognormal is clearly different: skewness and a series of far out are obvious. The Cauchy distribution stands out by its extreme far-out observations.

Exercise 1.23	
	Modify Example 1.13 so that the plots are comparable: adjust the location so that the medians are at the same height. Adjust the scales so that the inter-quartile ranges have same length.

Theorem 1.1 provides a possibility to determine confidence intervals for quantiles that are valid, independent of the underlying distribution.

To get an upper bound for the p-quantile x_p of a continuous distribution function by an order statistic $X_{(k:n)}$ with confidence level $1-\alpha$ we look for

$$\min_k : P(X_{(k:n)} \geq x_p) \geq 1-\alpha.$$

But $X_{(k:n)} \geq x_p \iff F(X_{(k:n)}) \geq p$ and by Theorem 1.1 we have

$$P(X_{(k:n)} \geq x_p) = 1 - F_{beta}(p; k, n-k+1).$$

So we can determine $\min_k$ directly using the beta distribution, or we use the relation with the binomial distribution and calculate k as

$$\min_k : P_{bin}(X \leq k-1; n, p) \geq 1-\alpha.$$

We will come back to this point in Section 3.5.2 (page 130).

Exercise 1.24	
	For continuous distributions and the median X_{med} we have $$P(X_i \geq X_{med}) = 0.5.$$ Hence we can find a k such that $$k = min\{k : P(X_{(k)} \leq X_{med}) < \alpha\}$$ and $X_{(k)}$ as an upper bound for the median with confidence level $1-\alpha$. Use this idea to construct a confidence interval for the median with confidence level $1-\alpha = 0.9$.
	Modify the box-and-whisker plot to show this interval. (cont.)→

Exercise 1.24	(cont.)
	Hint: You need the distribution function F_X, evaluated at the position marked by the order statistic $X_{(k)}$. The distributions of $F_X(X_{(k)})$ are discussed in Theorem 1.1.
	The box-and-whisker plot offers an option `notch = TRUE` to mark confidence intervals. Try to use the documentation to find out how a `notch` is calculated. Compare your confidence intervals with those marked using `notch`.
*	Use an analogous strategy to get a distribution-independent confidence interval for the inter-quartile range.
* * *	Augment the box-and-whisker plot so that it gives information about the scale in a way that is statistically reliable. *Hint:* Why is it not sufficient to mark confidence intervals for the quartiles?

1.5 R Complements

1.5.1 Random Numbers

If we had independent identical uniform distributed random numbers, we could generate random numbers with many other distributions. For example:

Lemma 1.12 *(Inversion method): If (U_i) is a sequence of independent identically $U[0,1]$ distributed random variables and F is a distribution function, then $(X_i) := (F^{-1}U_i)$ is a sequence of independent identically distributed random variables with distribution F.*

From an analytical point of view, this lemma is only usable if F^{-1} is known. Numerically, it has a much wider use. Instead of F^{-1} we often use only approximations, sometimes just an inversion table.

The inversion method is a method to use uniformly distributed random numbers to get random numbers with other distributions of interest. Other, sometimes much more efficient methods of obtaining random numbers with specific distributions are discussed in the literature on statistical simulation.

For a range of distributions, transformed random number generators are supplied with R. A list is given in Appendix A.21 (page A-228). For each distribution family there is a series of functions. The names for these functions are derived from the short name of the distribution. For the family xyz, `rxyz` is a function that generates random numbers. `dxyz` calculates the density or the point measure for this family. `pxyz` gives the distribution function and `qxyz` the quantiles.[3]

[3] Confusingly, with the notations that are common in statistics, we have $p_{xyz} \equiv$ `dxyz()` and $F_{xyz} \equiv$ `pxyz()`.

Distribution	*Random Numbers*	*Density*	*Distribution Function*	*Quantiles*
Binomial	`rbinom`	`dbinom`	`pbinom`	`qbinom`
Hypergeometric	`rhyper`	`dhyper`	`phyper`	`qhyper`
Poisson	`rpois`	`dpois`	`ppois`	`qpois`
Gauss (normal)	`rnorm`	`dnorm`	`pnorm`	`qnorm`
Exponential	`rexp`	`dexp`	`pexp`	`qexp`

Table 1.45 *Some selected distributions. For more distributions distributions see A.21 (page A-228).*

1.5.2 Graphical Comparisons

Differences between simple geometrical forms are perceived more easily than differences between general graphs of similar form. So it can be helpful to choose representations leading to simple forms such as straight lines. For example, to compare two distribution functions F, G, instead of comparing the function graphs we can consider the graph of

$$x \mapsto (F(x), G(x)).$$

This graph is called the *PP* ***plot*** or probability plot. If the distributions coincide, this graph shows a diagonal straight line. Deviations from the diagonal are perceived easily.

Alternatively, we can use the observation scale as a reference. So we consider the graph of

$$p \mapsto (F^{-1}(p), G^{-1}(p)).$$

This graph is called the *QQ* ***plot*** or ***quantile plot***. If the distributions coincide, this graph again shows a diagonal straight line.

For the special case of uniform distributions on $[0, 1]$ on this interval we have $x = F(x) = F^{-1}(x)$, that is, the *QQ* plot and the *PP* plot coincide. In this case, they both are just the graph of the distribution function. For non-uniform distributions in the *PP* plot the distribution functions are standardised to the probability scale $[0, 1]$, and in the *QQ* plot they are re-scaled to the observation scale.

Graphical methods are subject to the same critical differentiation as other statistical methods. The same type of criteria for appraisal applies as to formal methods. One approach, for example, is to ask for the discriminative power. As an example, let us study the normal approximation for $t(\nu)$-distributions. For one degree of freedom ($\nu = 1$) we have a Cauchy distribution that is very different from normal. Theory tells us that as the degree of freedoms ν increases, the $t(\nu)$-distributions converge to a normal distribution. We can get an impression of the discriminative power of different graphical methods by studying this transition:

Exercise 1.25	
	Generate a PP plot of the $t(\nu)$ distribution against the standard normal distribution in the range $0.01 \leq p \leq 0.99$ for $\nu = 1, 2, 3, \ldots$.
	Generate a QQ plot of the $t(\nu)$ distribution against the standard normal distribution in the range $-3 \leq x \leq 3$ for $\nu = 1, 2, 3, \ldots$.
	How large must ν be so that the t distribution is barely different from the normal distribution in these plots?
	How large must ν be so that the t distribution is barely different from the normal distribution if you compare the graphs of the distribution functions?

Of course aspects other than power may be considered when choosing a graphical method. So invariances or equivariances may be advantages of some method. For example, if the distributions are related by an affine transformation in the observation space, the QQ plot still shows a straight line; slope and intercept represent the affine transformation. This applies, for example, to the family of normal distributions: if F is the standard normal distribution $N(0, 1)$ and $G = N(\mu, \sigma^2)$, the QQ plot is a straight line with intercept μ and slope σ.

For the empirical distributions, Corollary 1.2 applies: instead of i/n we use a reference point that corrects for the skewness of the distribution, so that on average a straight line is generated. The quantile plot, using this correction for empirical distributions, is provided as function *qqplot()*. For the special case of the normal distribution there is a variation of *qqplot()* available as *qqnorm()* to compare an empirical distribution with the theoretical normal distribution.

By transforming to probability scale or observation scale the graphical procedures gain power. So, for example, using a sample size of $n = 50$ to tell the distribution function of the normal distribution from that of a uniform distribution often requires a well-trained observer. In the normal QQ plot, however, the uniform samples show up as markedly non-linear, whereas normal samples largely show linear pictures, and the difference becomes obvious.

To illustrate this, we generate random samples:

Input

```
unif50 <- runif(50)
unif100 <- runif(100)
norm50 <- rnorm(50)
norm100 <- rnorm(100)
lognorm50 <- exp( rnorm(50) )
lognorm100 <- exp( rnorm(100) )
```

We use these data sets to compare the visual information that is conveyed by the distribution function, in comparison to QQ plots. First we use these data sets to generate plots of the distribution functions.

Example 1.14: Distribution Functions (Various Distributions)

Input

```
oldpar <- par(mfrow = c(2, 3))

plot(ecdf(unif50), pch = "[")
plot(ecdf(norm50), pch = "[")
plot(ecdf(lognorm50), pch = "[")

plot(ecdf(unif100), pch = "[")
plot(ecdf(norm100), pch = "[")
plot(ecdf(lognorm100), pch = "[")

par(oldpar)
```

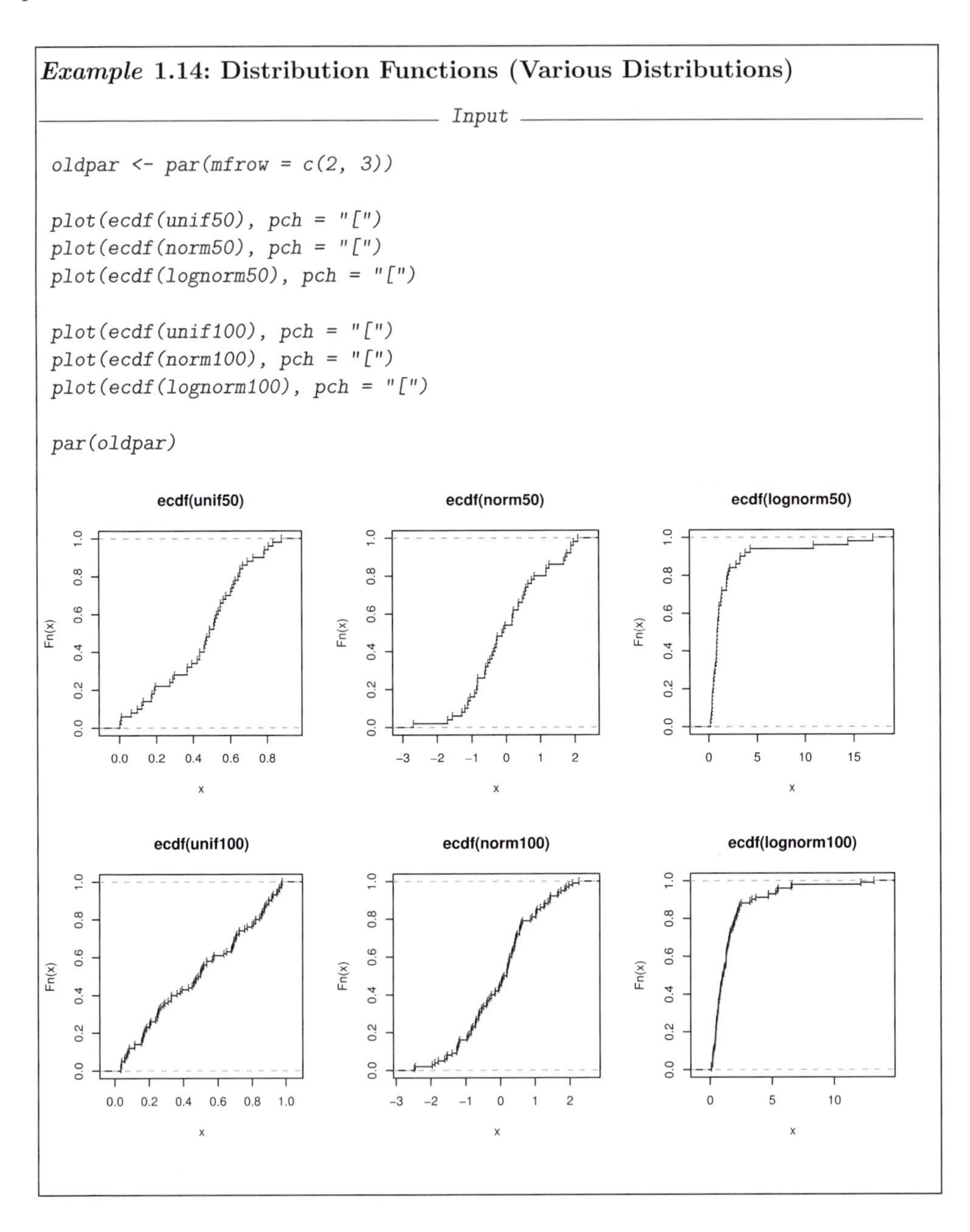

For comparison, here are the corresponding normal QQ plots for the same data:

***Example* 1.15: Normal *QQ* Plots (Various Distributions)**

Input

```
oldpar <- par(mfrow = c(2, 3))

qqnorm(unif50, main ="Normal Q-Q Plot\n unif50")
qqnorm(norm50, main = "Normal Q-Q\n norm50")
qqnorm(lognorm50, main = "Normal Q-Q\n lognorm50")

qqnorm(unif100, main = "Normal Q-Q\n unif100")
qqnorm(norm100, main = "Normal Q-Q\n norm100")
qqnorm(lognorm100, main = "Normal Q-Q\n lognorm100")

par(oldpar)
```

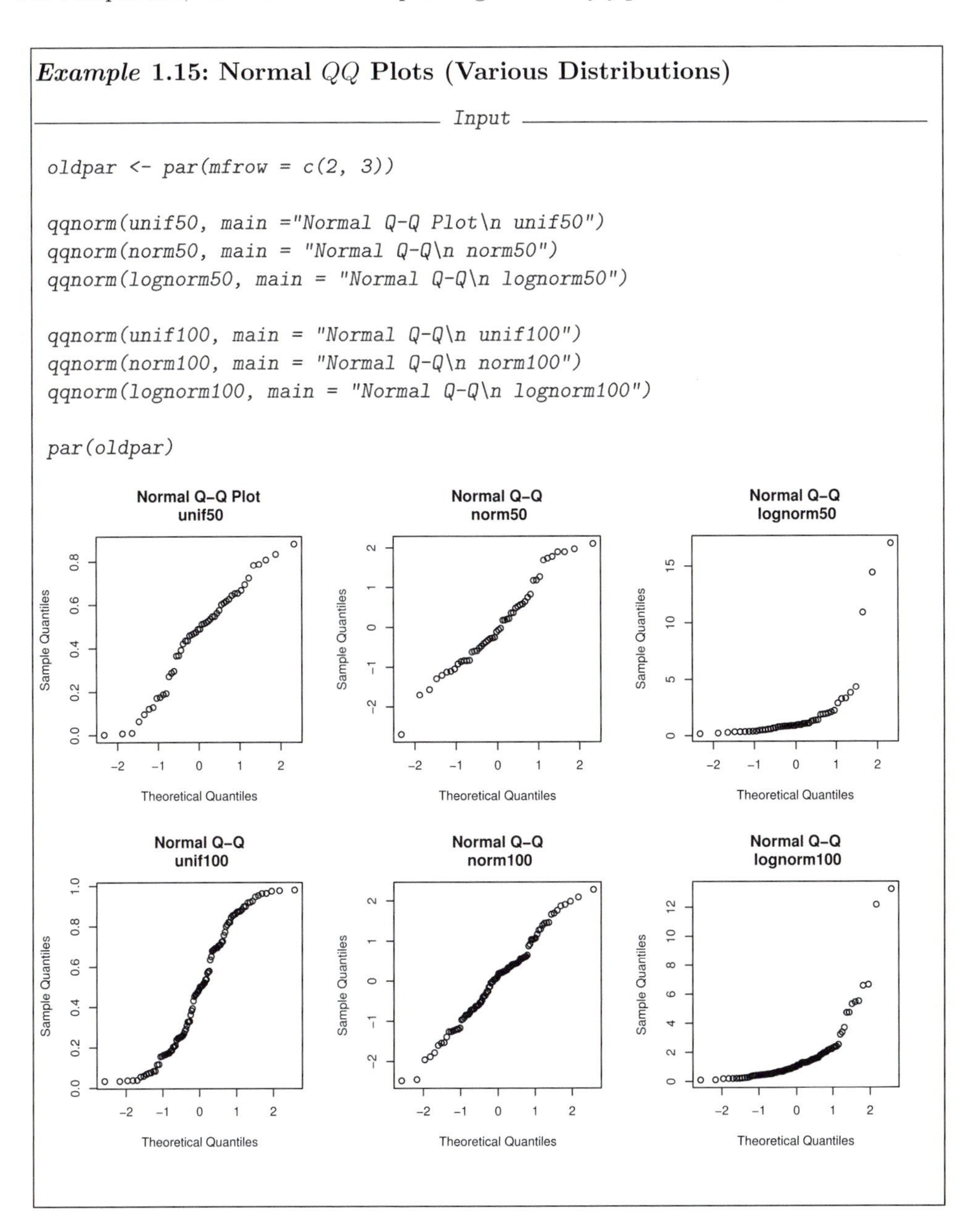

Exercise 1.26	
	Use PP plots instead of distribution functions to illustrate the χ^2- and Kolmogorov-Smirnov approximations.

Exercise 1.27	
	Use QQ plots instead of distribution functions. Can you add confidence regions to these plots with the help of the χ^2- resp. Kolmogorov-Smirnov statistics?

To get an impression of the fluctuation, we have to compare the empirical plots with typical plots of a model distribution. A plot matrix is a simple way to do this. Here is an example for the normal QQ plot, implemented as a function:

Input

```
qqnormx <- function(x, nrow = 5, ncol = 5, main = deparse(substitute(x))){
    oldpar <- par(mfrow = c(nrow, ncol))
    qqnorm(x, main = main)
    for (i in 1:(nrow*ncol-1))
     qqnorm(rnorm(length(x)), main = "N(0, 1)", xlab="", ylab="")
    par(oldpar)
}
```

In this example we used a `for` loop. Like all programming languages, R has control structures such as loops or conditional statements. In R, however, loops should be avoided if possible in favor of more efficient language constructs (see [25]). A summary of control structures in R can be found in Appendix A.14 (page A-213).

Exercise 1.28	
	Generate a matrix of dimensions $(nrow * ncol - 1), length(x)$ with random numbers and use `apply()` to avoid the loop. *Hint:* See Example 1.11 (page 23).

Deviations from a linear structure should be considered as fluctuations if they stay within the frame of the simulated examples. If the data set under investigation is too extreme in comparison with the simulated examples, this indicates a contradiction with the model assumptions. Because of space limitations of the printing area, we use only a small number of comparison plots here.

Example 1.16: Graphical Monte Carlo Test

Input

```
qqnormx(runif(100), nrow=4, ncol=4)
```

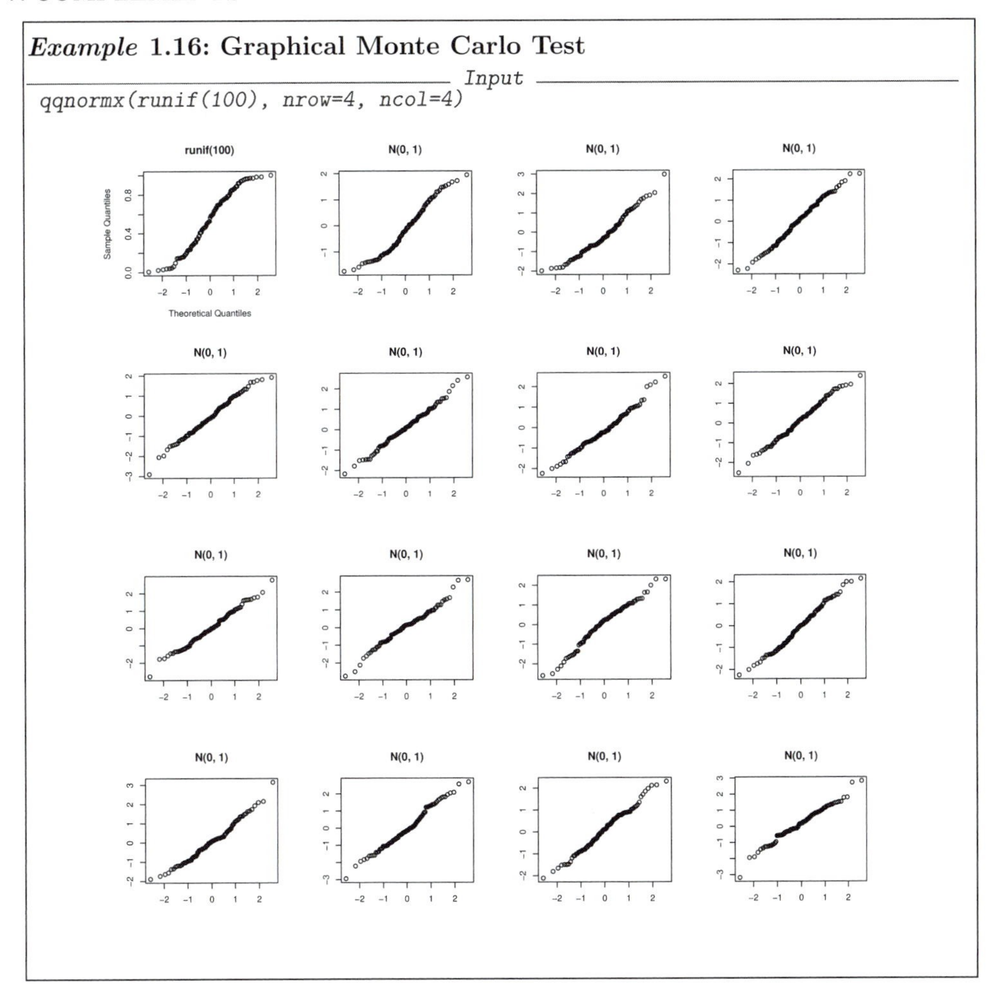

There are technical limitations. For printing in a book, a four-by-four scatterplot matrix is about the limit. Using a flip chart, or multiple pages, can extend the limit slightly. When working with an online display, a running sequence may be more convenient. A cheap solution is putting a test sample on the top left and running a series of simulations in the other quadrants of a two by two grid. Instead of `par`, we use `screen()` and `split.screen()` as a more convenient possibility of splitting the screen area.

```
split.screen(figs=c(2,2))
par(bg = "white")  # erase may not work if background is not set explicitely
screen(1)
qqnorm(runif(100), main="Uniform Sample")
for (i in (1:100)) {
    screen(2+(i %% 3), new=TRUE);
    qqnorm(rnorm(100), main="Normal Sample")
    #Sys.sleep(0.2)
}
```

If you try this, you will notice that on current computers the simulation runs too quickly, and you want more interactive control. This is an area of current development. Unfortunately, at present there are solutions for specific devices, but no matured portable solutions (see Appendix A.19.2 (page A-224)). You may try the code above and uncomment the call to `Sys.sleep()` if the function is available in your implementation. If you need a reliable, portable solution, you can open two devices, one for the test data set, and one for the simulations, and use function `devAskNewPage()` to force a console prompt whenever a new plot is to be drawn. As new solutions appear, they will be posted on the Web site <http://sintro.r-forge.r-project.org/>.

In the long run it is worth modifying the plot functions so that they give information about the fluctuation that is to be expected. In Example 1.12 we have constructed Monte Carlo bands for the distribution function. We can generalise this idea for the *PP* plot and the *QQ* plot. All that is needed is to transform the bands to the adequate scale used in the plot.

Exercise 1.29	
	Use `rnorm()` to generate with pseudo-random numbers for the normal distribution for sample size $n = 10, 20, 50, 100$. For each sample, generate a *PP* plot and a *QQ* plot, using the theoretical normal distribution as a reference.
	Add Monte Carlo bands from the envelope of 19 simulations. Instead of the uniform distribution, you have to use the normal distribution to generate the Monte Carlo bands. Then you have to represent the results in the coordinate system of the *QQ* plots, that is, the x axis represents the quantiles of the normal distribution. *Hint:* Inspect the source of `qqnorm()`.
*	The bands are initially bands for the standard normal distribution. Find bands adjusted in scale and location of the data at hand.

1.5.3 Complements: Functions

R commands can be grouped to functions. Functions may be parametrised by arguments. Functions provide a flexible way of code reusability.

Example of a function:

Input

```
ppdemo <- function (x, samps  =  19) {     # samps: nr of simulations
    y <- (1:length(x))/length(x)
    plot(sort(x), y, xlab = substitute(x), ylab = expression(F[n]),
        main = "Distribution Function with Monte Carlo Bands (unif.)",
        type = "s")
    mtext(paste(samps, "Monte Carlo Samples"), side = 3)
    samples <- matrix(runif(length(x)* samps),
        nrow = length(x), ncol = samps)
    samples <- apply(samples, 2, sort)
    envelope <- t(apply(samples, 1, range))
    lines(envelope[, 1], y, type = "s", col = "red");
    lines(envelope[, 2], y, type = "s", col = "red")
}
```

In `ppdemo()` we used the function `mtext()` to generate axis annotations.

Functions are called using the form ⟨name⟩(⟨actual argument list⟩).

***Example* 1.17: R Function Call**

Input

```
z100 <- runif(100)
ppdemo(z100)
```

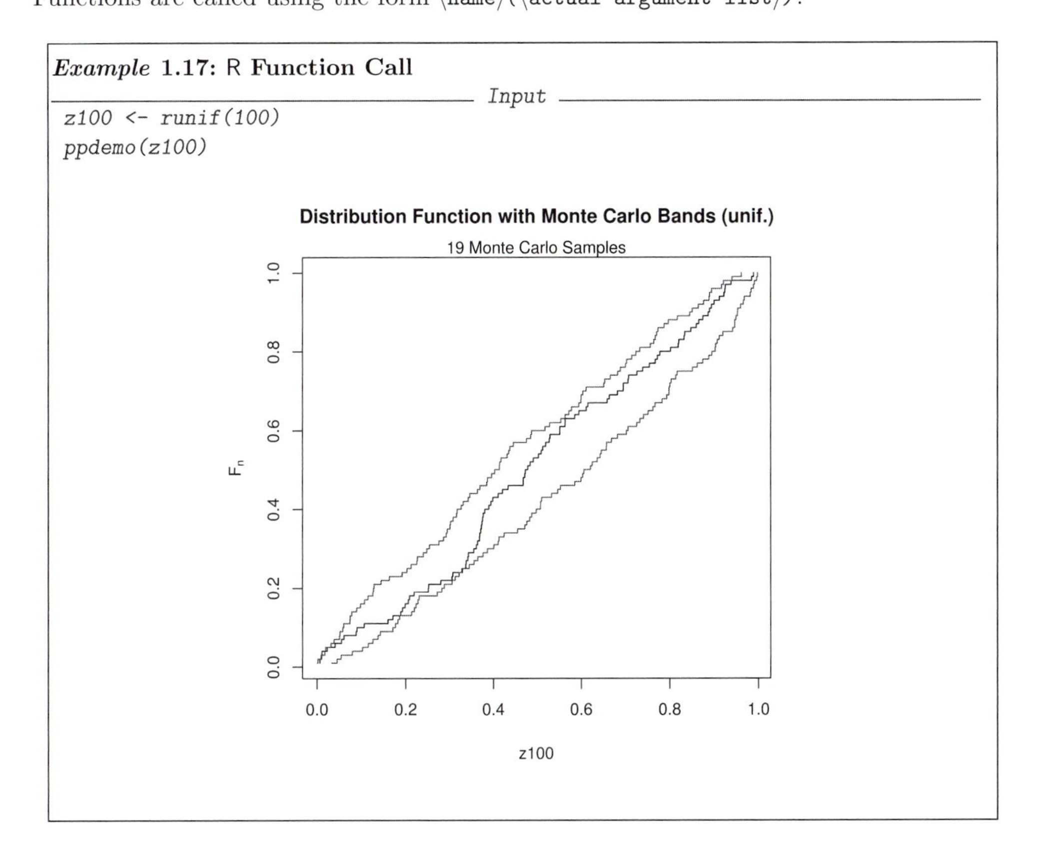

R *Function Declarations*	
Declarations	`function` (⟨`formal argument list`⟩) ⟨`expression`⟩ *Example:* `fak <- function(n) prod(1:n)`
Formal argument	⟨`argument name`⟩ ⟨`argument name`⟩ = ⟨`default value`⟩
Formal argument list	list of formal argument, separated by commas. *Examples:* `n, mean = 0, sd = 1`
...	Variable argument list. Variable argument lists can be propagated to embedded functions. *Example:* `mean.of.all <- function (...)mean(c(...))`
Function result	`return` ⟨`value`⟩ stops function evaluation and returns value.
	⟨`value`⟩ as last expression in a function declaration: returns value.
Function result	⟨`Variable`⟩`<<-`⟨`value`⟩ returns value. In general, assignments operate only on local copies of the variable. The assignment with `<<-`, however, looks for the target in the complete search chain.

R *Function Call*	
Function call	⟨`name`⟩`(`⟨`Supplied (actual) argument list`⟩`)` *Example:* `fak(3)`
Supplied argument list	Values are matched by position. Deviating from this, names can be used to control the matching. Initial parts of the names suffice (exception: after a variable argument list ..., names must be given completely). Function `missing()` can be used to check whether a corresponding actual argument is missing for a formal argument. *Syntax:* ⟨`list of values`⟩ ⟨`argument name`⟩ = ⟨`values`⟩ *Example:* `rnorm(10, sd = 2)`

If only the name of a function is given as input, the definition of the function is returned, that is, the function is listed. Example:

***Example* 1.18: R Function Listing**

Input

```
ppdemo
```

Output

```
function (x, samps  =  19) {      # samps: nr of simulations
    y <- (1:length(x))/length(x)
    plot(sort(x), y, xlab = substitute(x), ylab = expression(F[n]),
        main = "Distribution Function with Monte Carlo Bands (unif.)",
        type = "s")
    mtext(paste(samps, "Monte Carlo Samples"), side = 3)
    samples <- matrix(runif(length(x)* samps),
        nrow = length(x), ncol = samps)
    samples <- apply(samples, 2, sort)
    envelope <- t(apply(samples, 1, range))
    lines(envelope[, 1], y, type = "s", col = "red");
    lines(envelope[, 2], y, type = "s", col = "red")
}
```

Exercise 1.30	
	Rework your programming exercises and write reusable parts as functions.

Arguments for functions are passed by value. Each function receives a copy of the actual argument values upon call. This guarantees a safe programming environment. On the other side, this leads to memory requirements and a time penalty. In situations where the argument size is large or time requirements are critical this expenditure can be avoided by direct access to variables that are defined in the function environment. Techniques to achieve this are documented in [12].

Functions in R can be nested, that is, one "outer" function can contain other "inner" functions. These inner functions are not visible globally, but only in their containing outer functions.

Functions can have objects as results. An object is explicitly returned as a result by *return(obj)*. Results can be returned implicitly as well. If the end of a function is reached without calling *return()* the value of the expression evaluated last is returned.

Input

```
circlearea <- function( r) r^2 * pi
circlearea(1:4)
```

Output

```
[1]  3.141593 12.566371 28.274334 50.265482
```

Results can also be provided so that they are only passed upon request. We have encountered this technique when studying the histogram. The function call `hist(x)` does not return a result, but has the (intended) side effect of drawing a histogram. But if we use `hist()` in an expression, for example, in an assignment `xhist <- hist(x)`, we get a description of the histogram returned as the function value. To return results only upon request, instead of `return(obj)` the expression `invisible(obj)` is used.

Exercise 1.31	
	Write as functions: • A function `ehist` showing an augmented histogram. • A function `eecdf` showing the empirical distribution. • A function `eqqnorm` showing a *QQ* plot with the standard normal distribution as comparison. • A function `eboxplot` showing a box-and-whisker plot. and • A wrapper function `eplot` showing a plot matrix with these four plots. Your functions should call the standard functions (or modify them, if necessary) and guarantee that the plots have an adequate complete annotation.

While statements in R are processed stepwise and allow inspection of the results step by step, upon call of a function all statements of the function are executed as a block. This can pose a problem for error diagnosis. R provides possibilities for inspecting specific functions, in particular to allow stepwise processing of functions. For details see Appendix A.13 "Debugging and Profiling" (page A-211).

1.5.4 Complements: Enhancing Graphical Displays

So far the R graphics have been used in rudimentary form only. For serious work the graphics have to be enhanced to make their information identifiable and readable. This implies adding headers, captions, axis labels, etc. In R there are "high-level" and "low-level" graphic functions.

"High-level" functions generate a new display. Additionally, they provide possibilities to control general graphics parameters for use in this display.

The "low-level" functions add elements to an existing plot or modify it in some detail. So, for example, the function `legend()` can add legends inside a plot.

***Example* 1.19: Margin Text**

Input

```
plot(1:10, xlab = "xlab", ylab = "ylab", main = "main", sub = "sub")
mtext("mtext 1", side = 1, col = "blue")
mtext("mtext 2", side  = 2, col = "blue")
mtext("mtext 3", side  = 3, col = "blue")
mtext("mtext 4", side  = 4, col = "blue")
legend("topleft", legend = "topleft legend")
legend("center", legend = c("lty =1", "lty =2", "lty =3"),
    lty =  1:3, title = "center legend")
```

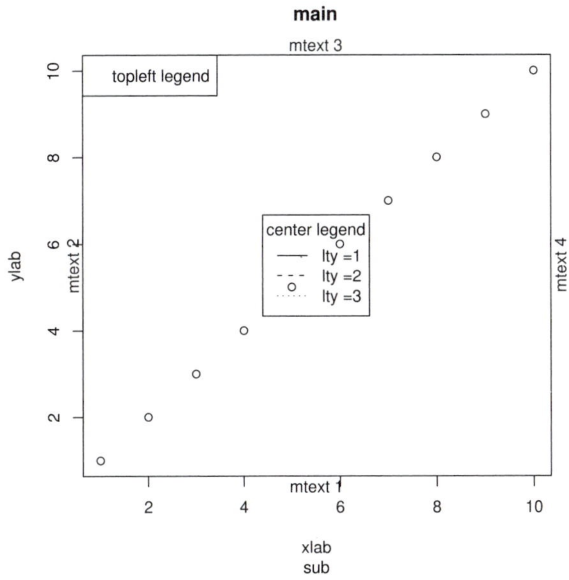

Exercise 1.32	
	Use `help(plot)` to inspect the possibilities of customising the plot function. Information on details of the parameters is only available if you use `help(plot.default)`. Modify your latest plot so it has a correct main title.

See also [30], Ch. 12.

For most plot types, annotation, choice of line width and symbols, and colours are at our disposal to enhance the plot. Sometimes, graphic enhancements may be necessary to obtain a useful plot at all. We give an illustration below, using colour options.

For this example, we need some background knowledge about colour specification. So far, we have specified colours by name, that is, we use a string such as `"red"` from a list of pre-specified names. The built-in list of colour names is available using `colours()`, but it is also possible to add user-defined colour names. The colours themselves are specified by a four-byte expression, with the bytes encoding the intensities for the red, green and blue channel, and the ***alpha channel*** that encodes opaqueness. For example, the hexadecimal value #000000FF would encode no colour intensity, but full opaqueness. So this would give black. The value #00000080 would still be black, but only 50% opaque. Drawing once with this colour will appear grey. But if drawings overlay, the colour will add up to full black.

Function *col2rgb()* is provided to translate from colour name to a vector of RGB values. Function *rgb()* translates from RGB and opaqueness to colour code.

Now let us look at the example. We construct a needle in the haystack. The haystack consists of n random points in the unit square. The needle is a small number $p \cdot n$ of these random points forced on the diagonal.

Example 1.20: Needle in the Haystack

Input

```
NeedleInTheHayStack <- function(nn, p=0.1, col="black", ... ) {
    oldpar <- par(mfrow=c(1,length(nn)))
    on.exit(par(oldpar))
    for (n in nn){
        nhay <- n-round(p*n); xhay <- runif(nhay);  yhay <- runif(nhay)
        needle <- runif(round(p*n))
        plot( x = c(xhay, needle), y = c(yhay,needle),
            main = paste("n = ", n, ", p = ", p, sep=""),
            cex.main=3.0,
            axes=FALSE, frame.plot=TRUE,
            xlab="", ylab="",
            col= col, ...)
        }
}
NeedleInTheHayStack( c(40, 200,1000, 5000, 25000) )
```


If n is very small, we have little chance of finding the needle in the haystack. As n increases, the structure becomes apparent. But if n gets large, the simple scatter plot is overloaded and we cannot access the information. This is a general problem, which is only delayed if we use a larger plotting area or smaller plot symbols. But we can enhance the plot using the alpha channel to make it useful again.[4]

[4] Visualising data sets for large sample sizes is a theme of its own. See [48].

***Example* 1.21: Needle in the Haystack**

Input
```
NeedleInTheHayStack( 25000, col = rgb(red=0, blue=0, green=0, alpha=0.1))
```

n = 25000, p = 0.1

The examples so far have used R's basic graphics system. Several other graphics systems are available to use with R. See, for example, the grid/lattice graphics system that we discuss in Section 4.1.

1.5.5 Complements: R *Internals*

A typical processing step in R handles a command in three sub-steps:

- `parse()` analyses a textual input and translates it into an internal representation as an R expression. R expressions are R objects of a special kind.
- `eval()` interprets this expression and evaluates it. The result is again an R object.
- `print()` displays the resulting object.

We explain these steps in detail below.

parse

The first step consists of two parts, a scanning that scans the input and splits it into tokens, and the parsing proper, which takes the tokens and tries to combine them into syntactically correct expressions. The function `parse()` combines both steps. `parse()` can work on local files or data streams as well as on external files denoted by an URL reference.

`parse()` does not resolve names of variables, but only generates an abstract parse tree. Variables in this tree can be resolved using `substitute()`.

As an inverse function `deparse()` is provided. A typical application is to decode the actual arguments of a function call (using `parse()`) and then to derive an informative annotation as in

```
xlab=deparse(substitute(x))
```

eval

The function `eval()` evaluates an R expression. For this step the references in expressions have to be translated into values, taking into account the actual environment. Since R is an interpreted system, environmental conditions can change. So the evaluation of one expression can lead to different results, depending on the environment.

Each function defines its own local environment. As functions can be nested, so can the environments. The (nested) environments define a ***search path***. Environments can also change as additional libraries are loaded into R. The actual local environment can be queried using `environment()`. With `search()` you get a list of the environments that are searched successively to resolve references. `ls()` gives a list of the objects in an environment.

R's extensibility implies the possibility that names can collide and the translation of names into actual values can become a problem. As a safeguard, R 2.x provides the possibility of collecting names in (guarded) name spaces. In most cases this is transparent for the user; the name resolution follows a search path that is determined by the chain of environments. To access an object in a specific name space explicitly, the name space can be specified. So, for example, `base::pi` can be used as an explicit name for the constant `pi` in the name space `base`, and this access will work even if the user has chosen to introduce some variable with the name `pi`, using some arbitrary value.

print

The function `print()` is implemented as a ***polymorphic*** function. To execute `print()` R evaluates the class of the objects to print and chooses an appropriate print method. Details are given in Section 2.6.5 (page 103).

Executing Files

The function `source()` is available to use a file as input for R. The file can be local, or specified by an URL reference. Conventionally for names of R source files, the suffix `.R` is used.

The function `Sweave()` [24] allows us to interweave documentation and commands. Conventionally, for names of `Sweave()` source files the suffix `.Rnw` is used. Details about the format are in the `Sweave()` documentation <http://www.statistik.lmu.de/~leisch/Sweave/>.

1.5.6 Complements: Packages

Functions, examples, data sets, etc. can be bundled in R packages following certain conventions. The conventions may change with different implementations. The conventions documented in R [35] should be used as a reference.[5] Some packages are part of R by default. Packages for special purposes can be found on the Internet, for example, using <http://www.cran.r-project.org/src/contrib/PACKAGES.html>.

Packages that are not bundled with R must be installed into the R system first. In general, system-specific commands are supplied for this installation. A more convenient way is to do the installation from inside R using `install.packages()`. If no special source is given, `install.packages` tries to access a prepared address (usually the CRAN address given above). But `install.packages` can load packages from any repository specified. In particular, using

[5] The use of terms in R may be slightly confusing. The current convention is to use `library` for the location and `package` for the content.

the form *install.packages(⟨package⟩, repos = NULL)*, ⟨package⟩ can be an access path for direct access on your local machine.

The function *update.packages()* compares installed versions with the recent state in the repository and updates the version installed locally if necessary.

Once a package is installed, it can be loaded using

```
library(pkgname)
```

After that, the objects (functions, data sets, ...) defined in the package are retrievable on the current search path and can be used directly in any expression.

Packages are released using

```
detach(pkgname)
```

After this, the package objects do not appear in the current search path any longer.

Technically, packages are built of directories, following the R conventions. Usually they are provided in packed and compressed form such as .tar.gz files. At the beginning, you will install pre-compiled binary versions of packages. If necessary, for example, to inspect the source code, the source version can be installed by the same functions.

For the organisation of your own work it is worth following the R conventions and to organise related parts as R packages. R provides several tools supporting package administration if you follow the conventions. The conventions and the tools available to work with packages are documented in [35]. For Unix/Linux/Mac OS X, the main tools are available as commands:

```
R CMD check <directory>#checks a directory for compliance with the R conventions
R CMD build <directory>#generates an R package
```

As a first step: The function *package.skeleton()* helps construct new packages. Besides generating a rudimentary package skeleton, it provides a help file documenting the next steps in building a loadable package.

Packages must have a file DESCRIPTION with specific information. The other components are optional. The details are documented in [35], and a prototype is generated by *package.skeleton()*.

Name	*Kind*	*Content*
DESCRIPTION	file	A source description following format conventions.
R	directory	R code. Files in this directory should be readable by *source()*. Recommended name suffix: .R
data	directory	Additional data. Files in this directory should be readable by *data()*. Recommended name suffixes and formats: .R for R code. Alternatively: .r .tab for tables. Alternatively: .txt, .csv

(cont.)→

Name	*Kind*	*Content*
		.RData for output of *save()*. Alternatively: .rda The directory should contain a file ***00Index*** with a short survey of the data sets included.
exec	directory	Additional executable files, for example, Perl- or shell scripts.
inst	directory	Contents are copied (recursively) to the target directory of the installation. In particular, this directory can contain a file CITATION, which is evaluated by the R function `citations()`.
man	directory	Documentation in the R documentation format (see [35], "Writing R extensions", available from <http://www.cran.r-project.org/>). Recommended name suffix: .Rd
src	directory	Fortran, C and other sources.
demo	directory	Executable examples. This directory should contain a file ***00Index*** with descriptions.

Exercise 1.33	
	Install the functions from Exercise 1.31 as a package. You can prepare the package with *package.skeleton()*, if you have already defined the functions. Load the package. Verify that you can still load the package with *library()* if you have restarted the R system. *Hint:* For an object x, the statement *prompt(x)* generates a skeleton upon which you can build a documentation for `x`.

1.6 Statistical Summary

In this chapter, our leading example was the statistical analysis of a (univariate) random sample. We used a model framework that is central to statistics. The values in the random sample are considered as realisations of a random variable, drawn from an underlying theoretical distribution. The aim of the statistical analysis was the inference from the empirical distribution of the random sample to the unknown underlying theoretical distribution. This inference can take two forms. We can compare the empirical distribution with some hypothetical distribution. This is the approach of classical statistics. Or we can attempt to extract features of the underlying distribution from the empirical distribution. This is the data analysis approach.

Both approaches are closely related. The essential tool for both was an analysis of the empirical distribution function.

1.7 Literature and Additional References

[35] R Development Core Team (2000–2008): Writing R Extensions.
See: `<http://www.r-project.org/manuals.html>`.

[46] Shorack, G. R.; Wellner, J. A.: *Empirical Processes with Applications to Statistics.*
Wiley, New York, 1986.

[11] Gänßler, P.; Stute, W.: *Wahrscheinlichkeitstheorie.*
Springer, Heidelberg, 1977.

[12] Gentleman, R.; Ihaka, R.: Lexical Scope and Statistical Computing.
Journal of Computational and Graphical Statistics 9 (2000) 491–508.

CHAPTER 2

Regression

2.1 General Regression Model

The (controlled) trial is a common paradigm from experimental sciences. Under experimental conditions x some measurement result y is recorded. This is composed of a systematic effect $m(x)$ and a measurement error ε:

$$y = m(x) + \varepsilon.$$

This is a particular view of the data. Instead of looking at the joint behaviour of x and y, an asymmetric view is taken: x is the "cause", y (or a change in y) is the effect. In this view, the "cause" x is considered as given or under direct control; y is influenced indirectly via the experimental conditions. Experimental conditions should be chosen so that ε, the measurement error, is kept as small as possible and vanishes on average: there should be no systematic error.

From a statistical point of view the main difference in the roles of x and y is that for y the stochastic behaviour is modelled using ε, while x is considered "given" and no stochastics is allotted for x in the model.

To get a manageable frame, we consider the case where x can be represented as a vector of real variables, $x \in \mathbb{R}^p$, and that the measurements are given as one-dimensional real values, $y \in \mathbb{R}$. In a stochastic model we can express the idea formally. One possible formalisation is to represent the measurement error ε as a random variable. Assuming additionally that the expected value of ε exists, the assumption that the measurement error vanishes on average can be formalised as $E(\varepsilon) = 0$.

To analyse the systematic effect m we consider a series of experiments. The index i, $i = 1, \ldots, n$, identifies an experiment in this series, and the model is

$$\begin{aligned} & y_i = m(x_i) + \varepsilon_i, \quad i = 1, \ldots, n \\ \text{with} \quad & x_i \in \mathbb{R}^p \\ & E(\varepsilon_i) = 0. \end{aligned}$$

Now the statistical problem is to estimate the function m from the measurement results y_i at measurement conditions x_i.

The keywords "curve estimation", "regression" or "ANOVA" lead to abundant statistical literature about this problem. The following discussion will only serve as an introduction.

We concentrate mainly on linear regression. This is a very simplified version of the general regression problem. But central aspects of regression can already be illustrated using the linear case. In practical applications, linear models are widely used, and general models are restricted to more sophisticated applications.

To have a unified terminology, we call y_i the ***response***, and the components of x_{ij} of x_i with $j = 1, \ldots, p$ are the ***regressors***. The function m is called the ***model function***.

Estimators are generally marked by a hat. So $\widehat{m}$ is an estimator for m.

We introduce two general terms. If we have any estimator, the evaluation at a point x results in values $\widehat{y} := \widehat{m}(x)$, the ***fit*** at point x. If we evaluate an estimator at an observation point, the fit $\widehat{m}(x_i)$ usually will differ from the observation y_i. The difference

$$R_{x_i}(y_i) := y_i - \widehat{m}(x_i)$$

is called the ***residual***. The residual can be seen as an estimator for the non-observable error term ε. Residuals do not exactly match the error terms. This would only be the case if the estimation were exact. The residuals are our only source of information about the error terms. They will be our starting point for inference about the error terms and they will provide the basis for judging the quality of estimators.

The estimation step is how this situation differs from the situation of independent, identically distributed observations that we discussed in Chapter 1. Typically, the estimator $\widehat{m}$ will combine information collected from all experiments in the experimental series. As a consequence, the fitted values, and hence the residuals, are random variables that are dependent on all experiments. So even if we could guarantee that the (non-observable) error terms in the original data were independent, the residuals will be dependent variables. We will have to invest additional effort beyond what has been discussed in Chapter 1 to compensate for this dependency.

2.2 Linear Model

We begin with the regression model, now in vector notation:[1]

$$\begin{aligned} Y &= m(X) + \varepsilon \\ &\quad Y \text{ observable random variable with values in } \mathbb{R}^n \\ &\quad X \in \mathbb{R}^{n \times p} \text{ known matrix} \\ &\quad E(\varepsilon) = 0 \text{ non-observable random variable.} \end{aligned} \tag{2.1}$$

Additionally, we assume that m is linear. Then there exists a vector $\beta \in \mathbb{R}^p$ (at least one), so that

$$m(X) = X\beta.$$

The regression problem is now reduced to the exercise to estimate β from the information (Y, X).

[1] We change conventions and notations whenever this change is helpful. The confusion is part of the conventions. In some common conventions, capital letters denote random variables. In other conventions, they denote functions, and in still others they denote vectors. We have to leave the resolution to the reader. We try to keep the convention to denote random variables by capital letters, and specific numerical outcomes are denoted by lower case letters. Greek letters are used for parameters. But the error term will still be ε.

The modified regression model

$$\begin{aligned} Y &= X\beta + \varepsilon \\ &\quad Y \text{ with values in } \mathbb{R}^n \\ &\quad X \in \mathbb{R}^{n\times p} \\ &\quad \beta \in \mathbb{R}^p \text{ unknown vector} \\ &\quad E(\varepsilon) = 0 \end{aligned} \tag{2.2}$$

is called a ***linear model*** or ***linear regression***. The matrix X, which combines the values of the regressors and hence the information about the experimental conditions, is called the ***design matrix*** of the model.

Example 2.1: (Simple Linear Regression) If the experimental condition is characterised by the value of one real variable, which links to the experimental result via a model function

$$m(x) = a + b \cdot x,$$

we can write a series of experiments with experimental result y_i, given experimental condition x_i, in coordinates as

$$y_i = a + b \cdot x_i + \varepsilon_i. \tag{2.3}$$

In matrix notation we can write the series of experiments as linear model

$$Y = \underbrace{\begin{pmatrix} 1 & x_1 \\ \vdots & \vdots \\ 1 & x_n \end{pmatrix}}_{X} \cdot \underbrace{\begin{pmatrix} a \\ b \end{pmatrix}}_{\beta} + \varepsilon. \tag{2.4}$$

Simple linear regression is one basic example for linear models. The other basic example is one-way classification.

Example 2.2: (One-Way Classification) For comparison of k treatments (in particular for the special case $k = 2$) we use indicator variables that are combined in a matrix. The indicator variable for treatment i is in column i. Usually we have repeated observations $j = 1, \dots, n_i$ under treatment i, giving a total of $n = \sum_{i=1}^{k} n_i$ observations. The model

$$Y = \underbrace{\begin{pmatrix} 1 & 0 & \dots & 0 \\ \vdots & \vdots & \vdots & \vdots \\ 1 & 0 & \dots & 0 \\ 0 & 1 & \dots & 0 \\ \vdots & \vdots & \vdots & \vdots \\ 0 & 1 & \dots & 0 \\ 0 & 0 & \dots & 1 \\ \vdots & \vdots & \vdots & \vdots \\ 0 & 0 & \dots & 1 \end{pmatrix}}_{X} \cdot \underbrace{\begin{pmatrix} \mu_1 \\ \mu_2 \\ \vdots \\ \mu_k \end{pmatrix}}_{\beta} + \varepsilon. \tag{2.5}$$

corresponds in coordinates to

$$y_{ij} = \mu_i + \varepsilon_{ij}, \qquad i = 1, \ldots, k, \qquad j = 1, \ldots, n_i. \tag{2.6}$$

This is the typical model to test the hypothesis "no difference" $\mu_1 = \ldots = \mu_k$ against the alternative that there is a difference in mean between the treatments.

The same relation can be represented if the measurement values are interpreted as a sum of a basic value μ_0 and an additional treatment effect $\mu'_i = \mu_i - \mu_0$. In coordinates, this reads

$$y_{ij} = \mu_0 + \mu'_i + \varepsilon_{ij}. \tag{2.7}$$

In matrix notation, this is

$$Y = \underbrace{\begin{pmatrix} 1 & 1 & 0 & \ldots & 0 \\ \vdots & \vdots & \vdots & \vdots & \\ 1 & 1 & 0 & \ldots & 0 \\ 1 & 0 & 1 & \ldots & 0 \\ \vdots & \vdots & \vdots & \vdots & \\ 1 & 0 & 1 & \ldots & 0 \\ 1 & 0 & 0 & \ldots & 1 \\ \vdots & \vdots & \vdots & \vdots & \\ 1 & 0 & 0 & \ldots & 1 \end{pmatrix}}_{X'} \cdot \underbrace{\begin{pmatrix} \mu_0 \\ \mu'_1 \\ \mu'_2 \\ \vdots \\ \mu'_k \end{pmatrix}}_{\beta'} + \varepsilon. \tag{2.8}$$

Example 2.2 illustrates that the representation of a problem as a linear model need not be unique. Equations (2.5) and (2.8) are equivalent representations and only the application background can decide which one is preferable.

For mathematical analysis, the design matrix X is an essential tool. For data analysis, we can use R to generate this matrix (implicitly) for us. R understands a special notation, the ***Wilkinson-Rogers notation***, for model description. With this notation we write

$$y \sim x.$$

The error term is not shown in this representation.

By default, a constant term is assumed. For the one-way classification we get model (2.8). If we do not want a constant term (in the regression case for a regression through the origin; in the one-way classification the model (2.5) where we do not include an overall mean), we have in coordinates

$$y_i = b \cdot x_i + \varepsilon_i.$$

In the Wilkinson-Rogers notation we have to set the constant term to zero explicitly:

$$y \sim 0 + x.$$

Additional regressors can be included using the operator +. So $y \sim u + v$ corresponds to the model (in coordinates)

$$y_i = a + b \cdot u_i + c \cdot v_i + \varepsilon_i.$$

We return to this notation in Section 2.2.5 (page 76) and Section 2.3 (page 79). A summary is given in Appendix A.18 (page A-221).

There is a rich literature on linear models. The book *The Theory of Linear Models* by Bent Jørgensen [22] deserves special recommendation. It covers most of the mathematical background of this chapter and contains numerous illustrative examples.

2.2.1 Factors

Using the notation for the description of designs and models, it is enough to give the rules how a model is built. A (case-oriented) description of a design can then be machine translated into a design matrix for a linear model. Sometimes the translation needs a little bit of coaching, as when a qualitative design variable is encoded using numeric code values. Consider, for example, a data set:

```
y <- c( 1.1, 1.2, 2.4, 2.3, 1.8, 1.9)
x <- c( 1, 1, 2, 2, 3, 3)
```

The vector x can mean a quantitative vector for the regression model

$$y_i \;=\; a + b\,x_i \;+\; \varepsilon_i$$

to be used as a regressor. Or it can be intended for one-way classification, the model of one-way analysis of variance,

$$y_{i\,x} \;=\; \mu + \alpha_x \;+\; \varepsilon_{i\,x},$$

where x is used as indicator for the treatment group. To mark the difference, vectors in R can be defined to be ***factors***. Numerical vectors that are not factors are treated as quantitative variables, as in the first example. Factors are treated as indicators and expanded into indicator variables when constructing the design matrix. So

```
y ~ x
```

results in the regression model, but

```
y ~ factor(x)
```

gives the variance model for the one-way layout.

Specifying a parameter `ordered = TRUE` when calling `factor()` marks the variable as ordered. The variable specified will be treated as on an ordinal scale:

```
y ~ factor(x, ordered = TRUE)
```

Without this specification, factors are considered to be on a categorical scale.

The values used for factors (both for categorical and ordered) may be mere names and will be substituted internally by numbers. If numerical values are used, they need not be in an ordered sequence. So

```
factor( c(2, 2, 5, 5, 4, 4) )
```

results in a vector having three factor values 1, 2, 3, with the names "2", "5" and "4". An example using a factor denoted by names would be, for example:

```
y ~ factor ( c("Tmt1", "Tmt1", "Tmt2", "Tmt2", "Tmt3", "Tmt3") )
```

The values that can be taken by a factor are called ***levels*** of the factor. This is an attribute of the factor that can be accessed with `levels()`, as, for example, in:

```
levels(factor( c(2, 2, 5, 5, 4, 4) ))
levels(factor( c("Tmt1", "Tmt1", "Tmt2", "Tmt2", "Tmt3", "Tmt3") ))
```

2.2.2 Least Squares Estimation

A first idea of estimation in a linear model can be gained from the following relation: given X, we have $E(Y) = X\beta$. As X is a matrix, we cannot simply solve this relation for β using a division by X. But we can expand the relation to $X^\top E(Y) = X^\top X\beta$. $X^\top X$ is a positive semi-definite symmetric matrix. If X has rank p, the full rank, this matrix is invertibe. In general at least a pseudo-inverse exists and we can calculate $(X^\top X)^- X^\top E(Y) = \beta$.[2] This equation motivates the following estimator:

$$\widehat{\beta} = (X^\top X)^- X^\top Y. \tag{2.9}$$

Using the model relation $Y = X\beta + \varepsilon$ 2.2 and $E(\varepsilon) = 0$, in the full rank case we get

$$E(\widehat{\beta}) = E\Big(\left(X^\top X\right)^- X^\top \left(X\beta + \varepsilon\right)\Big) = \beta, \tag{2.10}$$

so $\widehat{\beta}$ is an unbiased estimator for β. It is a topic in statistics lectures to discuss whether there are other qualities of this estimator. The Gauss-Markov theorem is a theorem from statistics characterising the estimator 2.2. We will come back to this estimator frequently. We give it a name for reference: the ***Gauss-Markov estimator***. In the case of a linear model, such as the regression model, this estimator has a series of optimality properties. For example, this estimator minimises the mean quadratic deviation, that is, it is a ***least squares estimator*** in this model.

The least squares estimator for linear models is implemented as function `lm()`.

We generate an example data set to be used for illustration.

Input

```
x <- 1:100
err <- rnorm(100, mean = 0, sd = 10)
y <- 2.5*x + err
```

For this example data set, we get the least squares estimator using

***Example* 2.3: Least Squares Estimator**

Input

```
lm(y ~ x)
```

Output

```
Call:
lm(formula = y ~ x)

Coefficients:
(Intercept)            x
     -1.416        2.502
```

[2] Here $X^\top$ means the transposed of the matrix X and $(X^\top X)^-$ denotes the (Penrose-Moore generalised) inverse of $(X^\top X)$.

Exercise 2.1	
	When we generated the data, we did not use a constant term. The model specified for estimation, however, did not exclude the constant term. Repeat the estimation using the model without a constant term. Compare the results.

The estimator $\widehat{\beta}$ immediately yields an estimation $\widehat{m}$ for the function m in our original model:

$$\widehat{m}(x) \;=\; x^\top \cdot \widehat{\beta}.$$

The evaluation at the measurement points results in the vector of the fitted values $\widehat{Y} = X\widehat{\beta}$.

In our example, the fit gives a regression line. Using *plot()* we can plot the data points. If we store the result of the regression, we can use it with *abline()* to add the regression line.

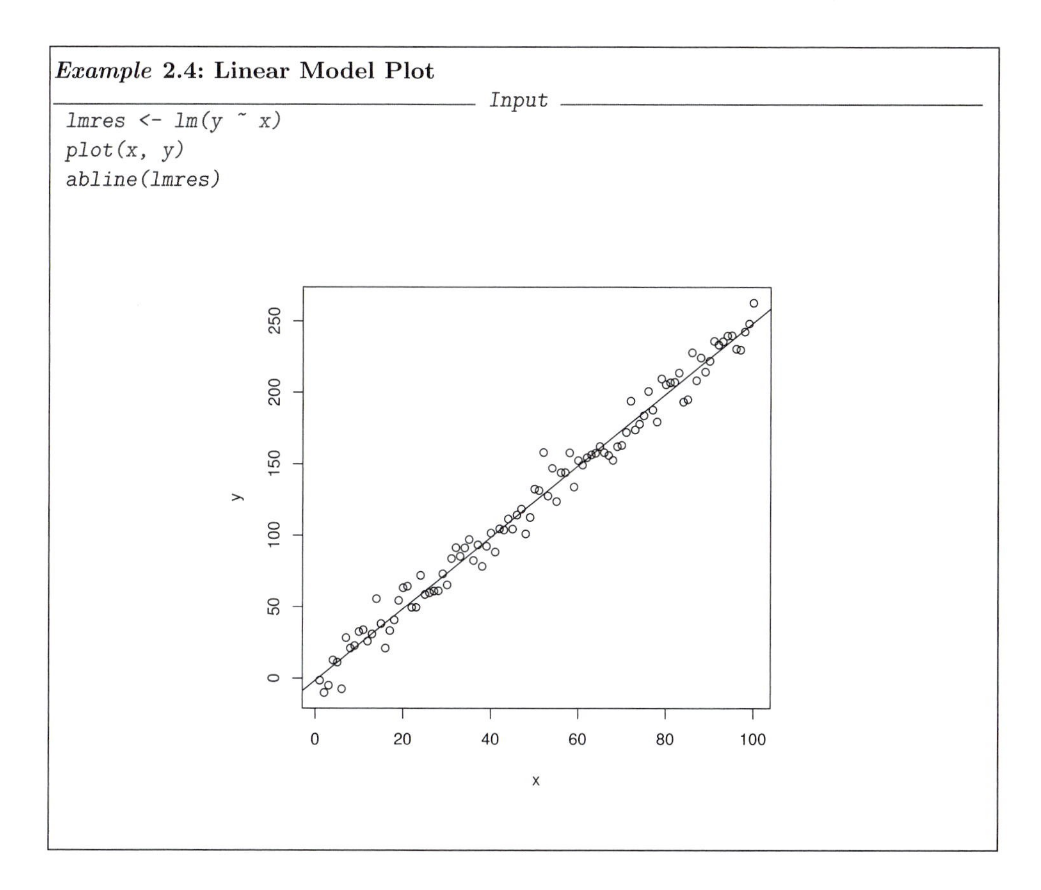

Example **2.4: Linear Model Plot**

Input

```
lmres <- lm(y ~ x)
plot(x, y)
abline(lmres)
```


Function *abline()* is a function to draw lines, using various parametrisations. For more information, see *help(abline)*.

Technically, we can apply the least squares estimation to any data set. The algorithm does not know whether the model assumptions apply, and it does not give us any information about the quality of the result. It is optimal, but optimality may not mean much if you are in dire straits. To judge the quality of the estimation, we need additional work.

The first step in this direction is to get information about the variance of the estimator. Equation 2.9 tells us that the estimator $\widehat{\beta}$ is a linear function of the observations Y. The matrix X is assumed to be known, hence the linear function is considered a known function. So the stochastic variation comes from the error terms contained in Y, and we have to reconstruct this.

Equation (2.9) tells us how to calculate the fit at the measurement points:

$$\widehat{Y} = X(X^\top X)^- X^\top \cdot Y. \tag{2.11}$$

The matrix

$$H := X(X^\top X)^- X^\top \tag{2.12}$$

is called the ***hat matrix***.[3] It is the main tool for analysing the Gauss-Markov estimator for a given design matrix X. The design matrix, and hence the hat matrix, depends only on the experimental conditions, not on the result of the experiment. The fit on the other side always refers to a specific outcome of the experiment, the random sample of observed values Y.

Writing Equation 2.11 as

$$\widehat{Y} = HY \tag{2.13}$$

highlights that the fit is a weighted average, a linear combination of the observations. Not all observations need to have the same weight. The coordinate representation

$$\widehat{Y}_i = H_{ii} \cdot Y_i + \sum_{j \neq i} H_{ij} \cdot Y_j \tag{2.14}$$

points to a potential problem. If the contribution H_{ii} is relatively large, the fit at data point i is dominated by the single observation Y_i. This can lead to gross errors if there is some problem at data point i. The diagonal elements H_{ii} are called ***leverages*** and are used as diagnostics for this kind of problem.

The linear model contains a term ε, representing the measurement error or the experimental fluctuation. We cannot observe this error directly. If we could, we would subtract it and get exact information about the model function. But since the error is not observable, we have to resort to indirect inference.

We already introduced the notion of residuals. To repeat: in general, the value of the random observation Y is different from the fit $\widehat{Y}$. The difference

$$R_X(Y) := Y - \widehat{Y}$$

is called ***residual***. The residual can be seen as an estimator for the non-observable error term ε. Residuals do not exactly match the error terms. This would only be the case if the estimation were exact. In our situation, the relation

$$\begin{aligned} R_X(Y) &= Y - \widehat{Y} \\ &= (I - H)Y \\ &= (I - H)(X\beta + \varepsilon) \\ &= (I - H)\varepsilon \end{aligned} \tag{2.15}$$

shows that the residuals are linear combinations of the error terms. We have to infer back from these linear combinations to the error term.

[3] It puts the hat on top of Y: $\widehat{Y} = H \cdot Y$.

If the variance of the error terms does exist, the variance matrix Σ of the error terms $Var(\varepsilon) = \Sigma$ determines the variance of the residuals:

$$\begin{aligned} Var(R_X(Y)) &= Var((I-H)\varepsilon) \\ &= (I-H)\Sigma(I-H)^\top. \end{aligned} \tag{2.16}$$

So far we have only presumed that there is no systematic error. This was formalised as the assumption

$$E(\varepsilon) = 0.$$

We speak of a ***simple linear model*** if we have additionally:

$$\begin{aligned} (\varepsilon_i)_{i=1,\dots,n} & \quad \text{are independent} \\ Var(\varepsilon_i) = \sigma^2 & \quad \text{for a } \sigma \text{ not depending on } i. \end{aligned}$$

For a linear model we try to estimate the parameter vector β. The variance structure of the vector of error terms introduces nuisance parameters, which complicate the estimation. For a simple linear model this nuisance reduces to just one unknown nuisance parameter σ. Equations like 2.16 can be simplified because now $\Sigma = \sigma^2 I$ and the parameter σ^2 can be pulled out from Formula 2.16. We can estimate this parameter from the residuals, because the ***residual variance***

$$s^2 := \frac{1}{n - Rk(X)} \sum_{i=1}^{n} (Y_i - \widehat{Y}_i)^2 \tag{2.17}$$

is an unbiased estimator for σ^2, where $Rk(X)$ is the rank of the matrix X. We write $\widehat{\sigma^2} := s^2$. (Taking the root is not a linear operation and does not preserve the expected value. The residual standard deviation $\sqrt{s^2}$ is not an unbiased estimator for σ.) Plugged into Equation (2.9), the residual variance estimator gives an estimator for the variance/covariance matrix of the estimator for β because in the simple model we have

$$Var\left(\widehat{\beta}\right) = \sigma^2 (X^\top X)^-, \tag{2.18}$$

which can be estimated by using the residual variance estimator as

$$\widehat{Var\left(\widehat{\beta}\right)} = s^2 (X^\top X)^-. \tag{2.19}$$

If in addition we can assume that the errors have a normal distribution, s^2 and $\widehat{\beta}$ are independent. We give a summary here, for simplicity for the full rank case where $Rk(X) = p$:

Theorem 2.1 *For a simple linear model with independent Gaussian errors and full rank $p = Rk(X)$ of the design matrix, the estimators $\widehat{\beta}$ and s^2 are stochastically independent:*

$$\widehat{\beta} \sim N_p(\beta, \sigma^2 X^\top X)^{-1}) \tag{2.20}$$

and

$$(n-p)\frac{s^2}{\sigma^2} \sim \chi^1_{n-p}. \tag{2.21}$$

If we standardise $\widehat{\beta}$ by $Var\left(\widehat{\beta}\right)$, each component has a t-distribution, that is, we can use t-tests for hypotheses such as $\beta_j = 0$.

The standard output in Example 2.3 shows only minimal information about the estimator. More information about the estimator, residuals and derived statistics are returned if we ask for a summary.

***Example* 2.5: Linear Model Summary**

Input

```
summary(lm( y ~ x))
```

Output

```
Call:
lm(formula = y ~ x)

Residuals:
     Min       1Q   Median       3Q      Max
-21.0700  -6.7568   0.4417   5.6749  29.3925

Coefficients:
            Estimate Std. Error t value Pr(>|t|)
(Intercept) -1.41598    1.92062  -0.737    0.463
x            2.50154    0.03302  75.762   <2e-16 ***
---
Signif. codes:  0 '***' 0.001 '**' 0.01 '*' 0.05 '.' 0.1 ' ' 1

Residual standard error: 9.531 on 98 degrees of freedom
Multiple R-squared: 0.9832,        Adjusted R-squared: 0.983
F-statistic:  5740 on 1 and 98 DF,  p-value: < 2.2e-16
```

Exercise 2.2	
	Analyse the output of `lm()` shown in Example 2.5. Which of the terms can you interpret? Write down your interpretations. For which terms do you need more information? Generate a commented version of the output.

In Section 2.3 (page 79) we will present the theoretical background needed to interpret the remaining terms.

A warning needs to be added here. R reports a t-test value and an error probability for each of the components of the parameter vector β. However, the estimation of the components and hence the derived t-tests are not independent. So to be on the safe side, you have to do a ***Bonferroni correction***, that is, if β has p components and you want to guarantee an error level of α, make sure that the nominal levels for your decision are at most α/p. This is a crude bound to keep you on the safe side. In special cases, it may be possible to have finer tools for simultaneous testing. Examples are in Section 2.4 (page 85).

2.2.3 Regression Diagnostics

Calling `lm()` always returns a result if it is appropriate for the data, but it will also return a linear result if the linear model is not adequate. We need additional diagnostics to tell us whether the model is reliable and usable.

Exercise 2.3	
	Let `yy <- 2.5*x +0.01 x^2 +  err` What are the results you get if you do a regression using the (incorrect) regression model $yy \sim x$? Do you get any hints that this model is not adequate?

The function `lm()` not only gives an estimation for the linear model, but also provides a series of diagnostics that can help to judge whether the model assumptions are acceptable. A representation using `plot()` shows some aspects.

***Example* 2.6: Linear Model Plot**

Input

```
plot(lm(y ~ x))
```

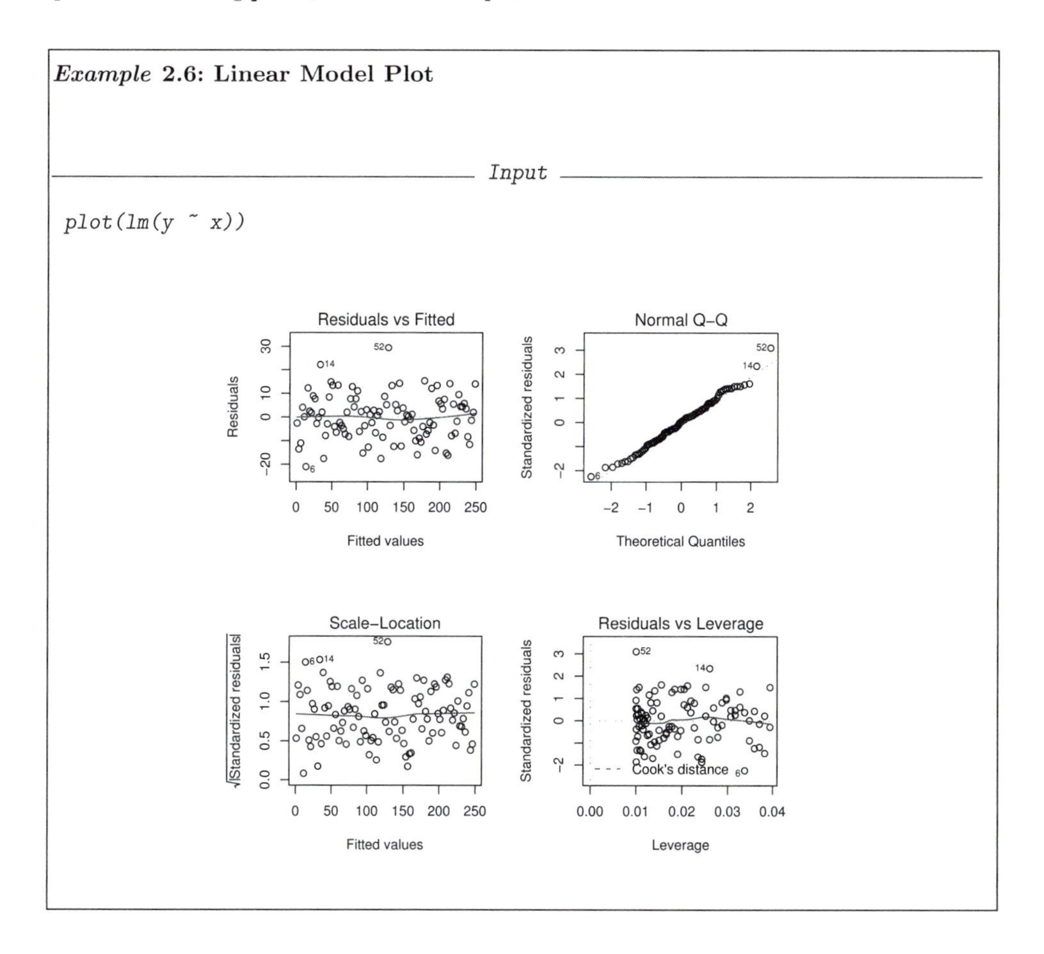

The top left plot shows the residuals against the fit. It gives a first survey.

The distribution of the fitted values depends on the design. Unless the design is homogeneous, you cannot expect the residual plot to be homogeneous.

The residuals should look approximately like a scatterplot of independent variables. The distribution of the residuals should not vary with the fit. If systematic structures show up in this plot, it is a warning that the model or the model assumptions may not be satisfied.

The previous discussion allows us to be more precise: the residuals should be linear combinations as in (2.15) of independent identically distributed variables. If the model assumptions are satisfied, the variance is given by (2.16).

In a one-dimensional situation, a plot of the residuals against the regressor would be sufficient. For p regressors, the graphical representation becomes difficult. The plot of the residuals against the fit, however, generalises to higher dimensions of the regressors.

If we start with distribution assumptions about the error terms, we can derive distribution properties of the estimator and of the residuals. The most powerful statements are possible if the error terms are independent identically distributed with a common normal distribution. In that case, the plot on the upper right should look approximately like a "normal probability plot" of normal random variates, where again "approximately" means: up to transformation with the matrix $I - H$. Using the empirical version of 2.16

$$\begin{aligned} Var\left(R_X(Y)\right) &= Var\left((I-H)\varepsilon\right) \\ &= (I-H)\widehat{\Sigma}(I-H)^\top \end{aligned} \tag{2.22}$$

and inverting it would give ***standardised residuals***. By convention, an approximation is taken as an algorithmic shortcut, giving

$$R_i^{(\text{std})} := \frac{R_i}{\sqrt{\widehat{\sigma^2}(1-H_{ii})}}. \tag{2.23}$$

The two remaining plots are special diagnostics for linear models (see `help(plot.lm)`). For example, the bottom-right plot is a scatterplot of the standardised residuals against leverage. Large standardised residuals are suspicious because they indicate a lack of fit. But small residuals may indicate a problem as well, in particular if they combine with a large leverage value. This hints at observations that may be outliers, possibly acting as leverage points with critical influence on the estimation. There is a rich literature on diagnostic plots which can be found using the keywords "residual analysis" or "regression diagnostics". As a concise textbook, see for example [55]. We will come back to this problem in Section 2.2.6 (page 77).

Exercise 2.4	
	Use `plot()` to inspect the results of Exercise 2.3. Does it give you indications that the linear model is not appropriate? Which indications?

`plot()` provides additional diagnostic plots for linear models. These must be requested explicitly using the parameter `which`.

help(lm)

lm	*Fitting Linear Models*

Description

`lm` is used to fit linear models. It can be used to carry out regression, single stratum analysis of variance and analysis of covariance (although `aov` may provide a more convenient interface for these).

Usage

```
lm(formula, data, subset, weights, na.action,
   method = "qr", model = TRUE, x = FALSE, y = FALSE, qr = TRUE,
   singular.ok = TRUE, contrasts = NULL, offset, ...)
```

Arguments

`formula` an object of class `"formula"` (or one that can be coerced to that class): a symbolic description of the model to be fitted. The details of model specification are given under 'Details'.

`data` an optional data frame, list or environment (or object coercible by `as.data.frame` to a data frame) containing the variables in the model. If not found in `data`, the variables are taken from `environment(formula)`, typically the environment from which `lm` is called.

`subset` an optional vector specifying a subset of observations to be used in the fitting process.

`weights` an optional vector of weights to be used in the fitting process. Should be `NULL` or a numeric vector. If non-NULL, weighted least squares is used with weights `weights` (that is, minimizing `sum(w*e^2)`); otherwise ordinary least squares is used.

`na.action` a function which indicates what should happen when the data contain `NA`s. The default is set by the `na.action` setting of `options`, and is `na.fail` if that is unset. The 'factory-fresh' default is `na.omit`. Another possible value is `NULL`, no action. Value `na.exclude` can be useful.

`method` the method to be used; for fitting, currently only `method = "qr"` is supported; `method = "model.frame"` returns the model frame (the same as with `model = TRUE`, see below).

`model, x, y, qr` logicals. If `TRUE` the corresponding components of the fit (the model frame, the model matrix, the response, the QR decomposition) are returned.

`singular.ok` logical. If `FALSE` (the default in S but not in R) a singular fit is an error.

`contrasts` an optional list. See the `contrasts.arg` of `model.matrix.default`.

`offset`	this can be used to specify an *a priori* known component to be included in the linear predictor during fitting. This should be `NULL` or a numeric vector of length either one or equal to the number of cases. One or more `offset` terms can be included in the formula instead or as well, and if both are specified their sum is used. See `model.offset`.
`...`	additional arguments to be passed to the low level regression fitting functions (see below).

Details

Models for `lm` are specified symbolically. A typical model has the form `response ~ terms` where `response` is the (numeric) response vector and `terms` is a series of terms which specifies a linear predictor for `response`. A terms specification of the form `first + second` indicates all the terms in `first` together with all the terms in `second` with duplicates removed. A specification of the form `first:second` indicates the set of terms obtained by taking the interactions of all terms in `first` with all terms in `second`. The specification `first*second` indicates the *cross* of `first` and `second`. This is the same as `first + second + first:second`.

If the formula includes an `offset`, this is evaluated and subtracted from the response.

If `response` is a matrix a linear model is fitted separately by least-squares to each column of the matrix.

See `model.matrix` for some further details. The terms in the formula will be re-ordered so that main effects come first, followed by the interactions, all second-order, all third-order and so on: to avoid this pass a `terms` object as the formula (see `aov` and `demo(glm.vr)` for an example).

A formula has an implied intercept term. To remove this use either `y ~ x - 1` or `y ~ 0 + x`. See `formula` for more details of allowed formulae.

`lm` calls the lower level functions `lm.fit`, etc, see below, for the actual numerical computations. For programming only, you may consider doing likewise.

All of `weights`, `subset` and `offset` are evaluated in the same way as variables in `formula`, that is first in `data` and then in the environment of `formula`.

Value

`lm` returns an object of `class "lm"` or for multiple responses of class `c("mlm", "lm")`.

The functions `summary` and `anova` are used to obtain and print a summary and analysis of variance table of the results. The generic accessor functions `coefficients`, `effects`, `fitted.values` and `residuals` extract various useful features of the value returned by `lm`.

An object of class `"lm"` is a list containing at least the following components:

`coefficients`	a named vector of coefficients
`residuals`	the residuals, that is response minus fitted values.
`fitted.values`	the fitted mean values.
`rank`	the numeric rank of the fitted linear model.
`weights`	(only for weighted fits) the specified weights.
`df.residual`	the residual degrees of freedom.
`call`	the matched call.

`terms`	the `terms` object used.
`contrasts`	(only where relevant) the contrasts used.
`xlevels`	(only where relevant) a record of the levels of the factors used in fitting.
`offset`	the offset used (missing if none were used).
`y`	if requested, the response used.
`x`	if requested, the model matrix used.
`model`	if requested (the default), the model frame used.
`na.action`	(where relevant) information returned by `model.frame` on the special handling of `NA`s.

In addition, non-null fits will have components `assign`, `effects` and (unless not requested) `qr` relating to the linear fit, for use by extractor functions such as `summary` and `effects`.

Using time series

Considerable care is needed when using `lm` with time series.

Unless `na.action = NULL`, the time series attributes are stripped from the variables before the regression is done. (This is necessary as omitting `NA`s would invalidate the time series attributes, and if `NA`s are omitted in the middle of the series the result would no longer be a regular time series.)

Even if the time series attributes are retained, they are not used to line up series, so that the time shift of a lagged or differenced regressor would be ignored. It is good practice to prepare a `data` argument by `ts.intersect(..., dframe = TRUE)`, then apply a suitable `na.action` to that data frame and call `lm` with `na.action = NULL` so that residuals and fitted values are time series.

Note

Offsets specified by `offset` will not be included in predictions by `predict.lm`, whereas those specified by an offset term in the formula will be.

Author(s)

The design was inspired by the S function of the same name described in Chambers (1992). The implementation of model formula by Ross Ihaka was based on Wilkinson & Rogers (1973).

References

Chambers, J. M. (1992) *Linear models.* Chapter 4 of *Statistical Models in S* eds J. M. Chambers and T. J. Hastie, Wadsworth & Brooks/Cole.

Wilkinson, G. N. and Rogers, C. E. (1973) Symbolic descriptions of factorial models for analysis of variance. *Applied Statistics*, **22, 392–9.**

See Also

summary.lm for summaries and anova.lm for the ANOVA table; aov for a different interface.

The generic functions coef, effects, residuals, fitted, vcov.

predict.lm (via predict) for prediction, including confidence and prediction intervals; confint for confidence intervals of *parameters*.

lm.influence for regression diagnostics, and glm for **generalized linear models.**

The underlying low level functions, lm.fit for plain, and lm.wfit for weighted regression fitting.

More lm() examples are available e.g., in anscombe, attitude, freeny, LifeCycleSaving longley, stackloss, swiss.

biglm **in package biglm for an alternative way to fit linear models to large datasets (especially those with many cases).**

Examples

```
require(graphics)

## Annette Dobson (1990) "An Introduction to Generalized Linear Models".
## Page 9: Plant Weight Data.
ctl <- c(4.17,5.58,5.18,6.11,4.50,4.61,5.17,4.53,5.33,5.14)
trt <- c(4.81,4.17,4.41,3.59,5.87,3.83,6.03,4.89,4.32,4.69)
group <- gl(2,10,20, labels=c("Ctl","Trt"))
weight <- c(ctl, trt)
anova(lm.D9 <- lm(weight ~ group))
summary(lm.D90 <- lm(weight ~ group - 1))# omitting intercept
summary(resid(lm.D9) - resid(lm.D90)) #- residuals almost identical

opar <- par(mfrow = c(2,2), oma = c(0, 0, 1.1, 0))
plot(lm.D9, las = 1)      # Residuals, Fitted, ...
par(opar)

## model frame :
stopifnot(identical(lm(weight ~ group, method = "model.frame"),
                    model.frame(lm.D9)))

### less simple examples in "See Also" above
```

To be added to the help information: in the formula notation, with two terms or lists of terms ***first*** and ***second***, ***first-second*** includes the variables indicated by the first term, but excludes those indicated by the second. For more information on the formula notation, see ***help(formula)***. A summary is given in Appendix A.18 (page A-221).

The hat matrix is a particuliarity of linear models. Fit and residuals, however, are general concepts and can be applied for all kind of estimations. Clients are often satisfied seeing a fit (or the estimation). For serious clients, and for statisticians, the residuals often contain more valuable information. They indicate what is not yet covered by the model or the estimation.

2.2.4 More Examples for Linear Models

The matrix X is called the ***design matrix*** of the model. It can be the matrix of row vectors x_i representing the measurement conditions for experiment i. But the design matrix is not restricted to this special case. The seemingly simple class of linear models comprises many important special cases. For example:

Simple Linear Regression:

$$y_i = a + b\, x_i + \varepsilon_i \qquad \text{with } x_i \in \mathbb{R}, a, b \in \mathbb{R}$$

can be written as a linear model,

$$X = (1\, x).$$

Here $1 = (1, \ldots, 1)^\top \in \mathbb{R}^n$.

Polynomial Regression:

$$y_i = a + b_1 x_i + b_2 x_i^2 + \ldots + b_k x_i^k + \varepsilon_i \qquad \text{with } x_i \in \mathbb{R}, a, b_j \in \mathbb{R}$$

can be written as a linear model

$$X = (1\ x\ x^2\ \ldots\ x^k)$$

where $x^j = (x_1^j \ldots x_n^j)^\top$.

By analogy, a multitude of models can be written as linear models, using other transformations.

Analysis of Variance: One-Way Layout

Observations are taken under m experimental conditions, with n_j observations taken under condition $j, j = 1, \ldots, m$. The measurement is the sum of a baseline effect μ, a contribution α_j that is, specific for condition j, and a measurement error

$$y_{ij} = \mu + \alpha_j + \varepsilon_{ij} \qquad \text{with } \mu, \alpha_j \in \mathbb{R},\ j = 1, \ldots, n_i.$$

This can be written as a linear model with $n = \sum n_j$ and

$$X = (1\ I_1\ \ldots\ I_m),$$

where I_j is an indicator variable marking group j.[4]

Analysis of Covariance

Like for the analysis of variance, differences between treatment groups are analysed. But in addition to the treatment, additional influence factors are considered and must be adjusted for. Under experimental condition j, the observation i in this group depends on additional influence factors x_{ij} for observation ij:

$$y_{ij} = \mu + \alpha_j + b\, x_{ij} + \varepsilon_{ij} \qquad \text{with } \mu, \alpha_j \in \mathbb{R}.$$

[4] For analysis of variance, the convention is to use the first index to mark the group, and the last index to count the observations per group. Index names are given in alphabetical sequence. Following this convention, in comparison to our notation, the roles of i and j are swapped.

2.2.5 Model Formulae

R allows us to specify models by giving the rules for building the design matrix. The syntax for writing these rules is sketched in the description of `lm()`, based on [56]. We will take a closer look at this formula notation. This kind of model specification not only applies to linear models, but can also be used for more general non-linear models. The model specification is stored as an attribute with the attribute name `formula`. It can be manipulated using `formula()`.

Examples

`y ~ 1 + x`	corresponds to $y_i = (1\ x_i)(\beta_1\ \beta_2)^\top + \varepsilon$.
`y ~ x`	short for $y \sim 1{+}x$ (a constant term is assumed implicitly).
`y ~ 0 + x`	corresponds to $y_i = x_i\beta + \varepsilon$.
`log(y) ~ x1 + x2`	corresponds to $\log(y_i) = (1\ \ x_{i1}\ \ x_{i2})(\beta_1\ \beta_2\ \beta_3)^\top + \varepsilon$ (a constant term is assumed implicitly).
`lm(y ~ poly(x, 4), data = Experiment)`	analyses the data set "Experiment" with a linear model for polynomial regression of degree 4 in x.

There are important special cases for factorial designs:

`y ~ A`	one-way analysis of variance with factor A.
`y ~ A + x`	analysis of covariance with factor A and regression covariable x.
`y ~ A + B`	two-factor crossed design with factors A and B without interaction.
`y ~ A * B`	two-factor crossed design with factors A and B and all interactions (combinations of the levels of A and B).
`y ~ A/B`	two-factor hierarchical layout with factor A and subfactor B.

A table of all operators for model specification formulae is in Appendix A.18 (page A-221).

Exercise 2.5	
	Write the four models from Section 2.2.4 using the R formula notation.
	For each of these models, generate an example data set by simulation, and apply `lm()` to the example. Compare the estimators returned by `lm()` with the parameters you have used in the simulations.

The model formula used by `lm()` is returned as an entry in the function result and recovered from there. From the formula, R implicitly generates a design matrix. Function `model.matrix()` can be used to inspect the design matrix.

Exercise 2.6	
	Generate three vectors of random variables with an $N(\mu_j, 1)$ distribution, $\mu_j = j,\ j = 1, 3, 9$, each of length 10, and combine these into a vector y. Generate a vector x with the values $j,\ j = 1, 3, 9$, each repeated 10 times. Calculate the Gauss-Markov estimator in the linear models `y~x` and `y~factor(x)`. Inspect the results as a table using `summary()` and graphically using `plot()`. Compare the results, and give a written report.

2.2.6 Gauss-Markov Estimator and Residuals

Let us take a closer look at the Gauss-Markov estimator. Knowledge from linear algebra, considerable thought or other sources tell us:

Remark 2.2

(1) The design matrix X defines a mapping $\mathbb{R}^p \to \mathbb{R}^n$ with $\beta \mapsto X\beta$. Let $\mathscr{M}_X$, $\mathscr{M}_X \subset \mathbb{R}^n$ be the image space of this mapping. $\mathscr{M}_X$ is the vector space generated by the column vectors from X.

(2) If the model assumptions are satisfied, $E(Y) \in \mathscr{M}_X$.

(3) $\widehat{Y} = \pi_{\mathscr{M}_X}(Y)$, where $\pi_{\mathscr{M}_X} : \mathbb{R}^n \to \mathscr{M}_X$ is the (Euclidean) orthogonal projection.

(4) In the full rank case, $\widehat{\beta} = arg\,min_\beta |Y - \widehat{Y}_\beta|^2$ where $\widehat{Y}_\beta = X\beta$.

The characterisation (3) of the Gauss-Markov estimator as an orthogonal projection often helps understanding. The fit is the orthogonal projection of the observation vector on the space of expected values of the model (which hence minimises the quadratic distance). This is the space spanned by the columns of the design matrix. The vector of residuals is the orthogonal complement.

In statistics, the estimator is analysed systematically, and the characterisation given above is just one starting point. Some properties of the estimator can be easily derived using knowledge from probability theory, such as the following lemma:

Theorem 2.3 *Let Z be a random variable with values in $\mathbb{R}^n$, with $N(0, \sigma^2 I_{n\times n})$ distribution, and let $\mathbb{R}^n = L_0 \oplus \ldots \oplus L_r$ be an orthogonal decomposition. Let $\pi_i = \pi L_i$ be the orthogonal projection onto L_i, $i = 0, \ldots, r$. Then the following holds:*

(i) $\pi_0(Z), \ldots, \pi_r(Z)$ are independent random variables with normal distributions.

(ii) $\frac{|\pi_i(Z)|^2}{\sigma^2} \sim \chi^2(\dim L_i)$ for $i = 0, \ldots, r$.

Proof. $\to$ probability theory. See, for example, [22], 2.5 Theorem 3. □

Using $\varepsilon = Y - X\beta$ allows us to derive the theoretical distributions for the estimator $\widehat{\beta}$ and the residuals $Y - \widehat{Y}$.

In particular, for simple linear models, the residual variance can be used to calculate the variance (resp. standard deviation) for each component $\widehat{\beta_k}$. The corresponding t statistics and the p-value for the test of the hypothesis $\widehat{\beta_k} = 0$ are given in the output of *summary()*.

Exercise 2.7	
	What is the distribution of $\|R_X(Y)\|^2 = \|Y - \widehat{Y}\|^2$, if ε has a $N(0, \sigma^2 I)$ distribution?

At first glance $|R_X(Y)|^2 = |Y - \widehat{Y}|^2$ seems an appropriate gauge to judge the quality of a model: small values indicate a good fit, large values indicate a poor fit. However, this has to be taken with caution. On the one hand, this value depends on linear scale factors. On the other hand, the dimension of the spaces involved has to be taken into account.

What happens if additional regressors are taken into the model? We have already seen that "linear" includes the possibility of modeling non-linear relations, for example, by taking transformed variables into the design matrix. The characterisation (3) in Remark 2.2 tells us that effectively only the vector space spanned by the design matrix is relevant. Here we can see limits for the Gauss-Markov estimator in linear models: if many transformed variables are taken into the model, or generally if the image space determined by the design matrix becomes too large, an over-fitting will result. In the extreme case we may get $\widehat{Y} = Y$. So all residuals are zero, but the estimation is not useful.

We use $|R_X(Y)|^2 / \dim(L_X)$, where L_X is the orthogonal complement of $\mathscr{M}_X$ in $\mathbb{R}^n$ (so $\dim(L_X) = n - \dim(\mathscr{M}_X)$) to compensate for the number of dimensions.

For diagnostic purposes, we go even further. We already have used a standardisation of the residuals in 2.23. If all assumptions are satisfied, this standardisation is sufficient. If we have possible outliers that work as leverage points and influence the regression critically, this influence on the estimation can lead to an over-fitting resulting in small residuals, effectively hiding the critical points. As a first precaution, 2.23 is modified. At any data point, the regression is calculated and the variance is estimated on the data set, excluding that data point. This gives a variant of the residuals called ***(externally) studentised residuals***. Standardised residuals are provided by `stdres()`; externally studentised residuals are available as `studres()`. Both are provided in library *MASS* [52].

Exercise 2.8	
	Modify the output of `plot.lm()` for the linear model so that instead of the Tukey-Anscombe plot the studentised residuals are plotted against the fit.
*	Enhance the *QQ*-Plot by Monte Carlo bands for independent normal errors. *Hint:* You cannot generate the bands directly from a normal distribution — you need the distribution of the residuals, not the distribution of the errors.

Exercise 2.9	
	Write a procedure that calculates the Gauss-Markov estimator for the simple linear regression $$y_i = a + bx_i + \varepsilon_i \quad \text{with } x_i \in \mathbb{R}, a, b \in \mathbb{R}$$ and shows four plots: • response against regressor, with estimated straight line • studentised residuals against fit • distribution function of the studentised residuals in a QQ plot with confidence bands • histogram of the studentised residuals

2.3 Variance Decomposition and Analysis of Variance

If a simple linear model with normal errors applies, t-tests are useful to solve one-dimensional problems (tests or confidence intervals for single parameters, point-wise confidence intervals). To solve simultaneous or higher-dimensional problems, we need other tools. Instead of differences or mean values that form the basis of t-tests, we use norm-based distances (as, for example, quadratic distances), which generalise to higher dimensions.

The interpretation of the Gauss-Markov estimator as an orthogonal projection (Remark 2.2, 3) shows a possibility to compare models: for design matrices X, X' with $\mathscr{M}_{X'} \subset \mathscr{M}_X$, we consider the decomposition $\mathbb{R}^n = L_0 \oplus \ldots \oplus L_r$ with $L_0 := \mathscr{M}_{X'}$, and the orthogonal complements $L_1 := \mathscr{M}_X \ominus \mathscr{M}_{X'}, L_2 := \mathbb{R}^n \ominus \mathscr{M}_X$. As above, π denotes the projection. As a statistics, we use the quotient of rescaled square norms

$$F := \frac{\frac{1}{\dim(L_1)} |\pi_{\mathscr{M}_X} Y - \pi_{\mathscr{M}_{X'}} Y|^2}{\frac{1}{\dim(L_2)} |Y - \pi_{\mathscr{M}_X Y}|^2}.$$

This statistic, the F statistic (in honour of R.A. Fisher) is the basis for the ***analysis of variance***, a classical strategy to compare models.

The idea generalises to chains of models. Let $\mathscr{M}_0 \subset \ldots \subset \mathscr{M}_r = \mathbb{R}^n$ be a chain of vector spaces, and $L_0 := \mathscr{M}_0, L_i := \mathscr{M}_i \ominus \mathscr{M}_{i-1}$ for $i = 1, \ldots, r$ the stepwise orthogonal complements. With the notation as above,

$$\frac{\frac{1}{\dim L_{i-1}} |\pi_{\mathscr{M}_i} Y - \pi_{\mathscr{M}_{i-1}} Y|^2}{\frac{1}{\dim L_i} |Y - \pi_{\mathscr{M}_i} Y|^2}$$

is a test statistic which can be used as a test for the model $\mathscr{M}_{i-1}$ in comparison to the refined $\mathscr{M}_i$. The same warning applies here as for the t-test. Unless you know exactly what you are doing, you have to do a ***Bonferrroni correction*** to adjust the nominal confidence level for multiple testing. Again, in special cases, it may be possible to have finer tools for simultaneous testing. Examples are in Section 2.4 (page 85).

Exercise 2.10	
	What is the distribution of F, if $E(Y) \in \mathscr{M}_{X'}$ applies and ε is distributed as $N(0, \sigma^2 I)$?

Exercise 2.11	
	Give an explicit formula for the F statistics for analysis of variance in the one-way layout $y_{ij} = \mu + \alpha_j + \varepsilon_{ij}$ in comparison to the homogeneous model $y_{ij} = \mu + \varepsilon_{ij}$.

The analysis of variance gives another representation and interpretation of linear models. For example, the regression result from 2.5 gives the following analysis of variance representation:

***Example* 2.7: Linear Model ANOVA Summary**

Input

```
summary(aov(lmres))
```

Output

```
          Df Sum Sq Mean Sq F value    Pr(>F)
x          1 521421  521421  5739.8 < 2.2e-16 ***
Residuals 98   8903      91
---
Signif. codes:  0 '***' 0.001 '**' 0.01 '*' 0.05 '.' 0.1 ' ' 1
```

Exercise 2.12	
	Analyse the output of `lm()` shown in Example 2.5 (page 68). Which terms can you interpret now? Give a written report. For which terms do you need more information?

One more item you find in the output is labelled "R-squared". The term that is given here is an estimator for the fraction of $Var(Y)$ which is explained by the model:

$$R^2 = \frac{mss}{mss + rss}$$

with $mss := \frac{1}{n}\sum(\widehat{Y}_i - \overline{\widehat{Y}})^2$ and $rss := \frac{1}{n}\sum(R_X(Y)_i - \overline{R_X(Y)})^2$. The notation R^2 stems from simple linear regression. In that context, conventionally the correlation $Cor(X, Y)$ is denoted by R, and $R^2 = Cor(X, Y)^2$. R^2 does not take into account the number of estimated parameters.

As a consequence, it can easily be too optimistic. The term "adjusted R-squared" is re-weighted to take into account the degrees of freedom.

---

help(anova)

`anova` *Anova Tables*

Description

Compute analysis of variance (or deviance) tables for one or more fitted model objects.

Usage

```
anova(object, ...)
```

Arguments

`object`	an object containing the results returned by a model fitting function (e.g., `lm` or `glm`).
`...`	additional objects of the same type.

Value

This (generic) function returns an object of class `anova`. These objects represent analysis-of-variance and analysis-of-deviance tables. When given a single argument it produces a table which tests whether the model terms are significant.

When given a sequence of objects, `anova` tests the models against one another in the order specified.

The print method for `anova` objects prints tables in a 'pretty' form.

Warning

The comparison between two or more models will only be valid if they are fitted to the same dataset. This may be a problem if there are missing values and R's default of `na.action = na.omit` is used.

References

Chambers, J. M. and Hastie, T. J. (1992) *Statistical Models in S*, Wadsworth & Brooks/-Cole.

See Also

`coefficients`, `effects`, `fitted.values`, `residuals`, `summary`, `drop1`, `add1`.

Models for the analysis of variance can be specified as descriptive rules. The syntax used is the same we have used for regression, that is, the Wilkinson-Rogers notation. If terms on the right-hand side of the model description are factors, `lm()` will automatically use a variance analysis model instead of a regression model.

The model description determines the linear spaces that contain the expected values for the model and its sub-models. The analysis of variance is not necessarily determined uniquely by the model description. The spaces may allow for various orthogonal decompositions (for example, depending on the order in which sub-models are considered). Moreover, the specification of the factors for a design determines a system of generators for the spaces, but these need not be linearly independent. Even in cases where they form a vector space basis, the factors need not be orthogonal.

This holds for all linear models. In regression, linear dependency is more an exception. With factorial designs, dependency is a frequent phenomenon. One-way analysis, written in coordinates, may illustrate this problem: with

$$y_{ij} = \mu + \alpha_j + \varepsilon_{ij} \quad \text{with } \mu, \alpha_j \in \mathbb{R}.$$

For $n_j > 0$ the decomposition into μ and α_j is not unique. The underlying reason is that the global factor μ defines the vector space spanned by the constant vector 1, and this is contained in the space spanned by the group indicators.

The model formula defines a design matrix X and hence a model space. An additional matrix C is used to extend the matrix and to specify a variance decomposition uniquely. The effective design matrix is then $[1 \;\; X \;\; C]$. C is called the ***contrast matrix***. The functions for analysis of variance like `lm()` or `aov()` allow for an explicit specification of contrasts.

The function `anova()` operates like a special formatting of the output and is used analogous to `summary()`, for example, using the form `anova(lm())`.

Exercise 2.13	**One-Way Anova**
*	Write a function `oneway()` which takes a data table as an argument and performs a one-way analysis of variance as a test on difference between the columns.
*	Enhance `oneway()` by adding the necessary diagnostic plots. Which diagnostics are necessary?

Exercise 2.14	**Kiwi Hopp**
	The industrial enterprise Kiwi Inc.[5] wants to develop a new helicopter for the market. The helicopter design is rated by the time it stays in air before it touches ground[6] from a fixed starting height (ca. 2m). Figure 2.1, page 84, shows a design drawing. What are the factors that can affect the variability of the flight (sink) time? What are the factors that can affect the mean flight duration?
	(cont.)→

[5] Following an idea of Alan Lee, Univ. Auckland, New Zealand.
[6] Kiwis cannot fly.

<table>
<tr><th>Exercise 2.14</th><th>Kiwi Hopp (cont.)</th></tr>
<tr><td></td><td>Perform 30 test flights with a prototype and measure the time in 1/100s. (You will have to cooperate in pairs to carry out the measurements.) Would you consider the recorded times as normally distributed?

The requirement is that the mean flight duration reaches at least 2.4s. Does the prototype satisfy the requirement?</td></tr>
<tr><td></td><td>Your task is to select a design for production. The variants under discussion are:

rotor width 45mm
rotor width 35mm
rotor width 45mm with an additional fold for stabilisation
rotor width 35mm with an additional fold for stabilisation

Your budget allows for about 40 test flights. (If you need more test flights, you should give good arguments for this.) Build 4 prototypes, perform the test flights and record the times. Find the design that achieves maximum flight duration. Generate a report. The report should contain the following details:

• a list of the observed data and a description of the experimental procedure
• suitable plots of the data for each of the designs
• an analysis of variance
• a clear summary of your conclusions

Additional hints: Randomise the sequence of your experiments. Reduce the variation by providing uniform conditions for the experiment (same height, same launch technique, etc.).</td></tr>
<tr><td></td><td>The fold will result in additional production cost. Give an estimate of the gain that can be achieved by this additional cost.</td></tr>
</table>

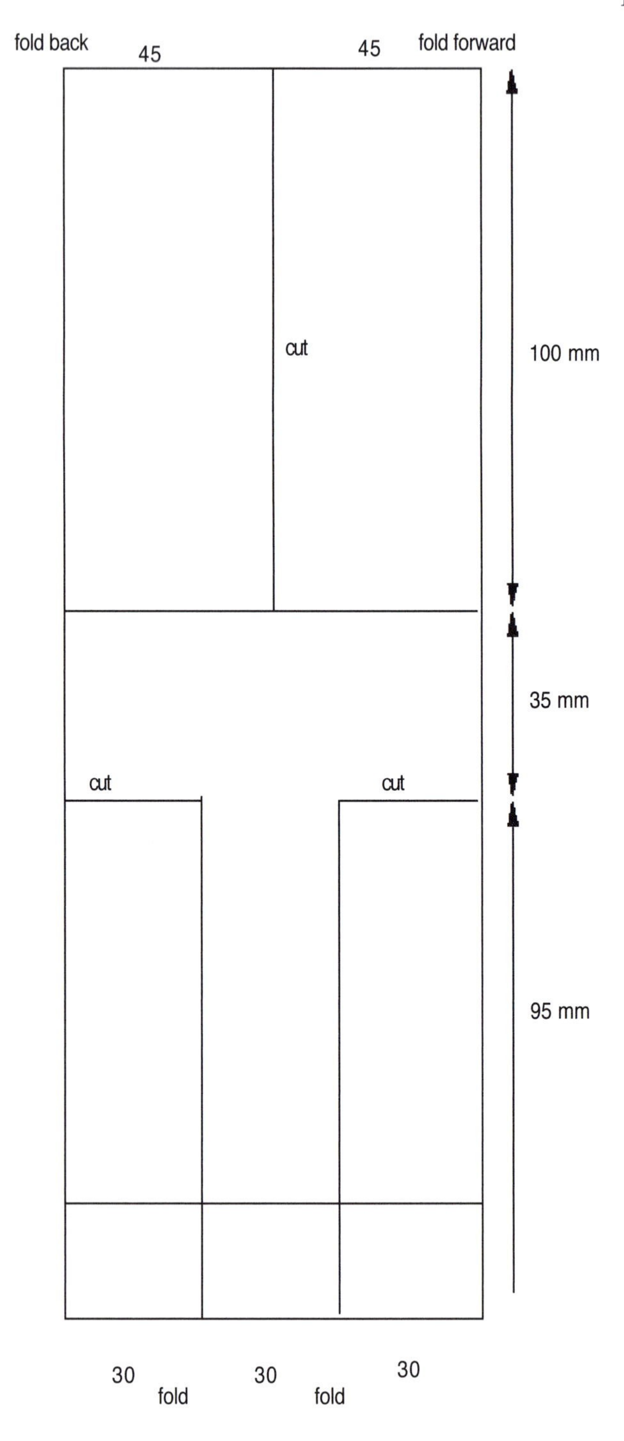

Figure 2.1 *KiwiHopp*

2.4 Simultaneous Inference

2.4.1 Scheffé's Confidence Bands

In principle, the least-squares estimator estimates all components of the parameter vector simultaneously. On the other hand, the optimality statements in the Gauss-Markov theorem refer only to one-dimensional linear statistics. But multivariate confidence statements are possible. The confidence set for confidence level $1 - \alpha$ derived from the F distribution has the form

$$\{\widehat{\beta} \in \mathbb{R}^k : (\sum_{j=1}^{k} (\widehat{\beta}_j - \beta_j)^2 \|x_j\|^2 / k) / \widehat{\sigma^2} \le F_{1-\alpha}(k, n-k)\}.$$

This confidence set is an ellipsoid. Instead of thinking of a quadratic form, we can see the ellipsoid as the area delimited by all tangents of the ellipsoid. This translates the (single) quadratic condition on the points in the confidence set into (infinitely many) linear conditions. This geometric relation is the core of the following theorem:

Theorem 2.4 *Let $\mathscr{L} \subset \mathbb{R}^k$ be a linear subspace of dimension d; $EY = Xb$ with $Rk(X) = p$, $p < n$. Then*

$$P\{\ell^t \beta \in \ell^t \widehat{\beta} \pm (dF^{\alpha}_{d,n-\alpha})^{1/2} s(\ell^t (X^t X)^{-1} \ell)^{1/2} \; \forall \ell \in \mathscr{L}\} = (1 - \alpha).$$

Proof. [28], 2.2, p. 48 □

This is a simultaneous confidence set for all linear combinations from $\mathscr{L}$. Translated into a test, this result gives a simultaneous test for all linear hypotheses from $\mathscr{L}$. In the special case $d = 1$ this simultaneous Scheffé test reduces to the usual F-test. In general, it is not possible to evaluate several tests on the same data material without deflating the confidence level. The F-test is an exception. After a global F-test it is possible to test these linear combinations or contrasts individually, without violating the level of the test.

In the case of simple linear regression, we get a confidence ellipsoid in the parameter space, that is, the space with coordinates *intercept*, *slope*. If we look at the lines or surfaces defined by the parameter values in this confidence ellipsoid, we get a hyperboloid as the confidence set for the regression in the regressor/response space.

In the regressor/response space, that is, the space of experimental conditions and observations, the interest is often not in a confidence set for the regression, but instead in a prediction region for observations to come. To get a tolerance area, the stochastic variation of the error term affecting additional observations must be added to the estimation uncertainty. So the tolerance area is increased in comparison to the estimation hyperboloid for the regression. Confidence sets for the regression and tolerance regions for observation can be calculated using the function `predict()`. The following figure shows both regions. The function `predict()` is a generic function. For linear models, it calls `predict.lm()`. `predict()` allows us to specify new supporting points where the fit is calculated using the estimated model parameters. Variables are matched here by name. `newdata` has to be a data frame with component names corresponding to those of the original variables. We prepare a data set as an example.

Input

```
n <- 100
sigma <- 1
x <- (1:n)/n-0.5
err <- rnorm(n)
y <- 2.5 * x + sigma*err

lmxy <- lm(y ~ x)
```

To get better control of the graphics, we calculate the plot limits and

Input

```
plotlim <- function(x){
    xlim <- range(x)
    # check implementation of plot. is this needed?
    del <- xlim[2]-xlim[1]
    if (del>0)
        xlim <- xlim+c(-0.1*del, 0.1*del)
    else xlim <- xlim+c(-0.1, 0.1)
    return(xlim)
}

xlim <- plotlim(x)
ylim <- plotlim(y)
```

Next, we generate the supporting points.

Input

```
newx <- data.frame(x = seq(xlim[1], xlim[2], 1/(2*n)))
```

For these nodes, we calculate the confidence bands and draw them. We apply `predict` to calclate the bands. `matplot` is used to do the actual drawing.

Input

```
pred.w.plim <- predict(lmxy, newdata = newx, interval = "prediction")
pred.w.clim <- predict(lmxy, newdata = newx, interval = "confidence")
```

***Example* 2.8: Linear Model Confidence Bands**

Input

```
plot(x, y, xlim = xlim, ylim = ylim)
abline(lmxy)
matplot(newx$x,
    cbind(pred.w.clim[, -1], pred.w.plim[, -1]),
    lty = c(2, 2, 6, 6),
    col = c(2, 2, 4, 4),
    type = "l", add = TRUE)
title(main = "Simultaneous Confidence")
legend("topleft",
    lty = c(2, 6),
    legend = c("confidence", "prediction"),
    col = c(2, 4),
    inset = 0.05, bty = "n")
```

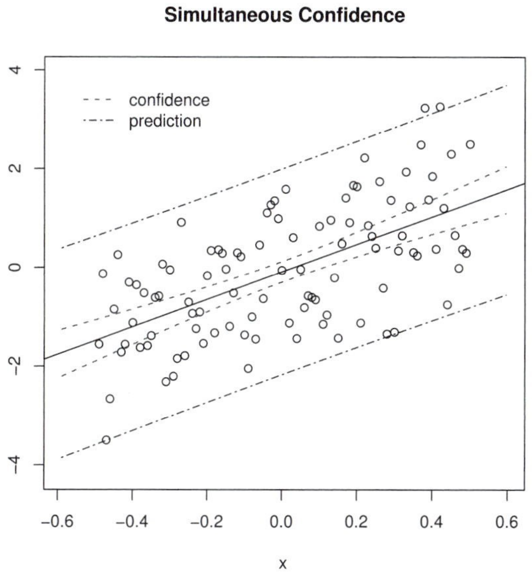

2.4.2 Tukey's Confidence Intervals

There is more than one way to allot a given error probability to multiple tests. Scheffé's confidence bands are one solution, and geometrically they are nearest to the idea underlying F-tests. But other ideas, using other rationales, are competing. Geometrically, the confidence ellipsoid used in Scheffé's approach is delimited by its (infinitely many) tangential spaces. Translated into tests this means that infinitely many linear tests are performed simultaneously. In many applications it is possible to formulate more specific questions. For example, to compare k treatments with treatment effects β_i, $i = 1, \ldots, k$ we may be interested only in the hypothesis $\beta_i - \beta_{i'} = 0$. This reduced question can be formulated in the framework of linear models. When

specified, they can lead to more powerful tests than the omnibus simultaneous tests. Families of hypotheses of this kind can again be specified as ***contrasts***. R supports the use of contrasts for the analysis of variance.

Case Study: Titre Plates

A typical tool in biology and medicine are titre plates, which are used, for example, in cell culture experiments. The plate has small wells in a rectangular grid. Some substrate can be applied to the plate as a whole. Then on each well, a test substance can be applied. Typically, this is done at once using multi pipettes for a row or a column (Figure 2.2).

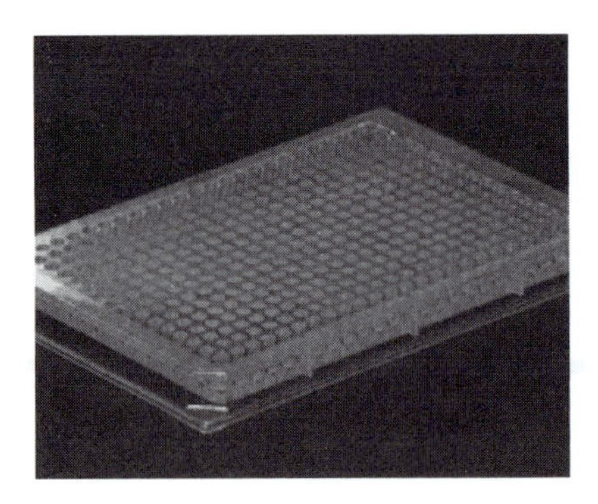

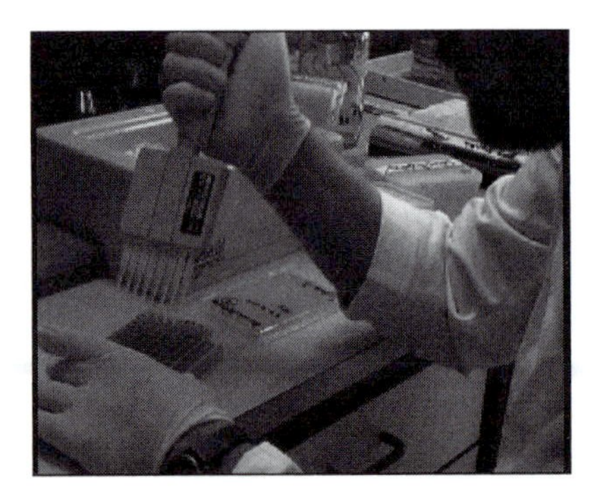

Figure 2.2 *Titre plates. With multi pipettes, substances can be at once applied to a row or a column.*

The experiments are often done in series. We will use just one plate from a series as our data example. If the data files are installed, for example in a subdirectory `../data/`, you can use

Input
```
p35 <- read.delim("../data/p35.tab")
```

R allows access to external data. So if you do not have a local copy of the data set, but access to the internet, you can use an online access such as

```
p35 <- read.delim("http//sintro.rforge.r-project.org/data/p35.tab")
```

In this experiment, data had been recorded in matrix form. For the analysis with `lm()` we have to transform the data from a matrix form to a long form that has a treatment label in one column. The column *H* of the plate in our experiment contains no treatment, but serves only for comparison of plates in the series for quality control.

Input
```
s35 <- stack(p35[,3:9])                                    # ignore column H
s35 <- data.frame(y=s35$values,
     Tmt = s35$ind,
     Lane = rep( 1:12, length.out = dim(s35)[1]) )  # rename
lmres <- lm(y ~ 0+ Tmt, data= s35)   # we do not want an overall mean
```

The summary as a linear model gives t-tests for the individual coefficients.

Input

```
summary(lmres)
```

Output

```
Call:
lm(formula = y ~ 0 + Tmt, data = s35)

Residuals:
      Min        1Q    Median        3Q       Max
-0.084833 -0.016354  0.009125  0.022729  0.073083

Coefficients:
     Estimate Std. Error t value Pr(>|t|)
TmtA  0.19383    0.01035   18.73   <2e-16 ***
TmtB  0.24892    0.01035   24.06   <2e-16 ***
TmtC  0.23783    0.01035   22.99   <2e-16 ***
TmtD  0.24117    0.01035   23.31   <2e-16 ***
TmtE  0.24392    0.01035   23.57   <2e-16 ***
TmtF  0.23558    0.01035   22.77   <2e-16 ***
TmtG  0.22367    0.01035   21.62   <2e-16 ***
---
Signif. codes:  0 '***' 0.001 '**' 0.01 '*' 0.05 '.' 0.1 ' ' 1

Residual standard error: 0.03584 on 77 degrees of freedom
Multiple R-squared: 0.9787,      Adjusted R-squared: 0.9768
F-statistic: 506.2 on 7 and 77 DF,  p-value: < 2.2e-16
```

For this example, tests for individual coefficients are not adequate. *anova()* gives a summary that is tailored to the needs of an analysis of variance.

Input

```
anova(lmres)
```

Output

```
Analysis of Variance Table

Response: y
          Df Sum Sq Mean Sq F value    Pr(>F)
Tmt        7 4.5513  0.6502  506.15 < 2.2e-16 ***
Residuals 77 0.0989  0.0013
---
Signif. codes:  0 '***' 0.001 '**' 0.01 '*' 0.05 '.' 0.1 ' ' 1
```

Provided the assumptions of the Gauss-linear model are satisfied, this summary says that the treatment effect is significant. The next question that arises immediately is which of the treatments differ significantly, that is, we are interested in the contrasts that describe the differences between treatments. Without violating the level of the test, we can use Tukey's approach to post-hoc tests. This is supported by the function *glht()* for tests of generalised linear hypotheses, which is provided in the package *multcomp* [15] for multiple testing.

Example **2.9: Tukey's Multiple Comparison**

Input

```
library(multcomp)
lhtres <- glht(lmres, linfct=mcp(Tmt="Tukey"))
summary(lhtres)   # multiple tests
```

Output

```
         Simultaneous Tests for General Linear Hypotheses

Multiple Comparisons of Means: Tukey Contrasts


Fit: lm(formula = y ~ 0 + Tmt, data = s35)

Linear Hypotheses:
            Estimate Std. Error t value p value
B - A == 0  0.055083   0.014632   3.765 0.00597 **
C - A == 0  0.044000   0.014632   3.007 0.05256 .
D - A == 0  0.047333   0.014632   3.235 0.02838 *
E - A == 0  0.050083   0.014632   3.423 0.01642 *
F - A == 0  0.041750   0.014632   2.853 0.07792 .
G - A == 0  0.029833   0.014632   2.039 0.39924
C - B == 0 -0.011083   0.014632  -0.757 0.98819
D - B == 0 -0.007750   0.014632  -0.530 0.99832
E - B == 0 -0.005000   0.014632  -0.342 0.99986
F - B == 0 -0.013333   0.014632  -0.911 0.96971
G - B == 0 -0.025250   0.014632  -1.726 0.60097
D - C == 0  0.003333   0.014632   0.228 0.99999
E - C == 0  0.006083   0.014632   0.416 0.99958
F - C == 0 -0.002250   0.014632  -0.154 1.00000
G - C == 0 -0.014167   0.014632  -0.968 0.95931
E - D == 0  0.002750   0.014632   0.188 1.00000
F - D == 0 -0.005583   0.014632  -0.382 0.99974
G - D == 0 -0.017500   0.014632  -1.196 0.89359
F - E == 0 -0.008333   0.014632  -0.570 0.99748
G - E == 0 -0.020250   0.014632  -1.384 0.80870
G - F == 0 -0.011917   0.014632  -0.814 0.98280
---
Signif. codes:  0 '***' 0.001 '**' 0.01 '*' 0.05 '.' 0.1 ' ' 1
(Adjusted p values reported -- single-step method)
```

Under the assumptions of the model the significance of the differences between A and B, the differences between A and D, and A and E are established at a level of 5%.

To check the assumptions, we can use the residuals, which can be inspected with `plot()`.

***Example* 2.10: Multiple Comparison Model Check**

Input

```
oldpar <- par(mfrow=c(2,2))
plot(lmres)
par(oldpar)
```

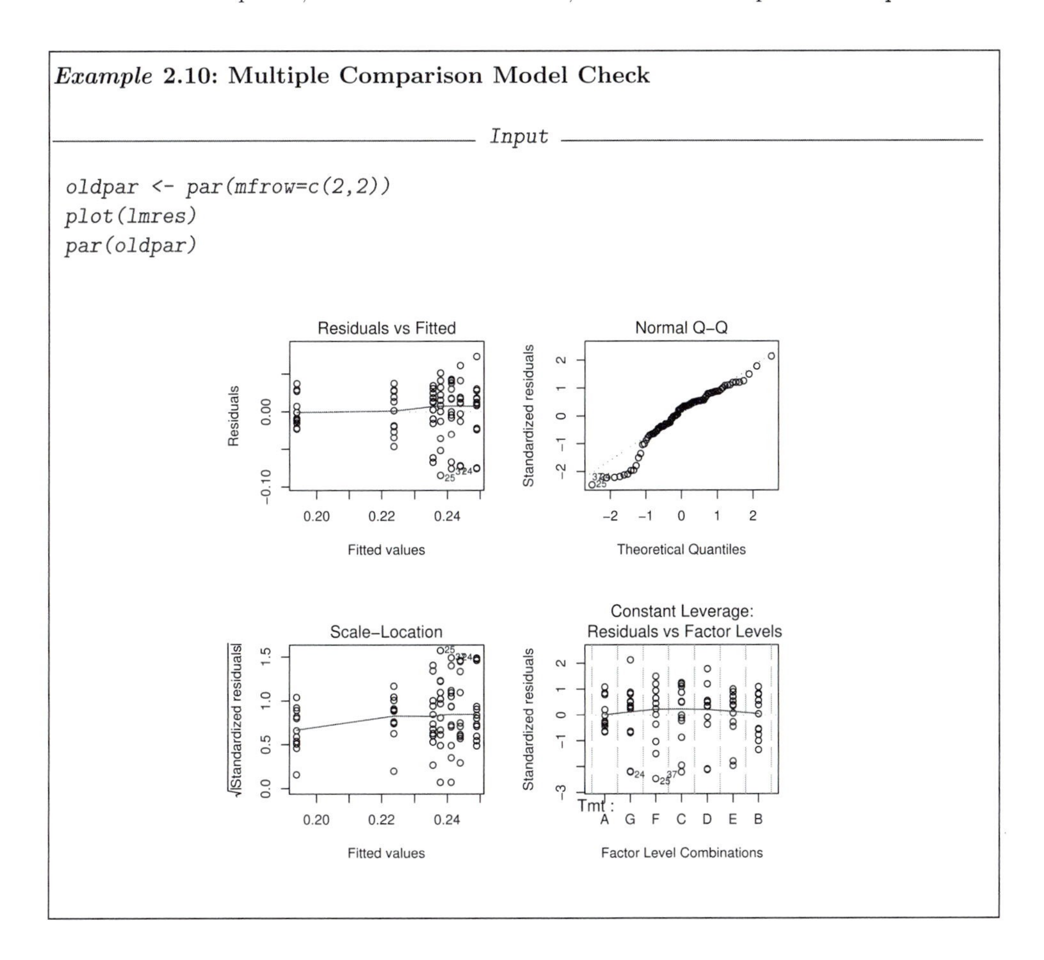

The distribution shows a clear deviation from the normal distribution, in particular at low values. We can inspect these selectively.

Example **2.11: Diagnostic Using Studentized Residuals**

Input

```
library(MASS)
s35$studres <- studres(lmres)
s35[s35$studres < -1,]
```

Output

```
       y Tmt Lane    studres
13 0.174   B    1 -2.239390
24 0.173   B   12 -2.271296
25 0.153   C    1 -2.559766
33 0.202   C    9 -1.044865
36 0.186   C   12 -1.523409
37 0.165   D    1 -2.279286
48 0.174   D   12 -1.994858
49 0.171   E    1 -2.175828
60 0.172   E   12 -2.144169
61 0.168   F    1 -2.007887
72 0.174   F   12 -1.821449
73 0.177   G    1 -1.367608
84 0.189   G   12 -1.010381
```

The pattern is conspicuous. Nearly all small values occur at the boundary of the plate in lane 1 or 12.

We could as well have detected this pattern by visual inspection. The built-in function `image()` would allow a graphical representation of the matrix, but it follows plot conventions, which place the origin at the bottom left corner, whereas we are used to viewing matrices as a table with an origin at the top left. So instead of using `image()`, we use a private modification, which follows the matrix conventions and gives a display that keeps the aspect ratio of the original matrix.

As is usual with R, either we find a package that serves our needs, or we customise the system. Our first step is to get a colour scale that highlights the extremes (and is perceptible even for most people with colour-impaired vision). Here is an example:

Input

```
blueyellow4.colors <-
function (n=100, rev=FALSE)
{
        if ((n <- as.integer(n[1])) > 0) {
                n<-n-1
                if (rev) {
                q<-((n:0)/n  -0.5) *2
                }
```

```
                else {
                q<-((0:n)/n  -0.5) *2
                }
                qq <- ((q*q*q*q*sign(q)+1)/2)
                q1<- 1- ((q*q*q*q+1)/2)
                rgb(qq+q1,qq+q1, 1-qq+q1)

            }
     else character(0)
}
```

This produces a colour ramp as in the following example:

***Example* 2.12: Custom Colour Palette**

Input

```
oldpar <- par(yaxt="n")
image(x=1:12, z=matrix(1:12,ncol=1),
    col=blueyellow4.colors(12),
    xlab="",  main="Colour Codes for Rank")
par(oldpar)
```

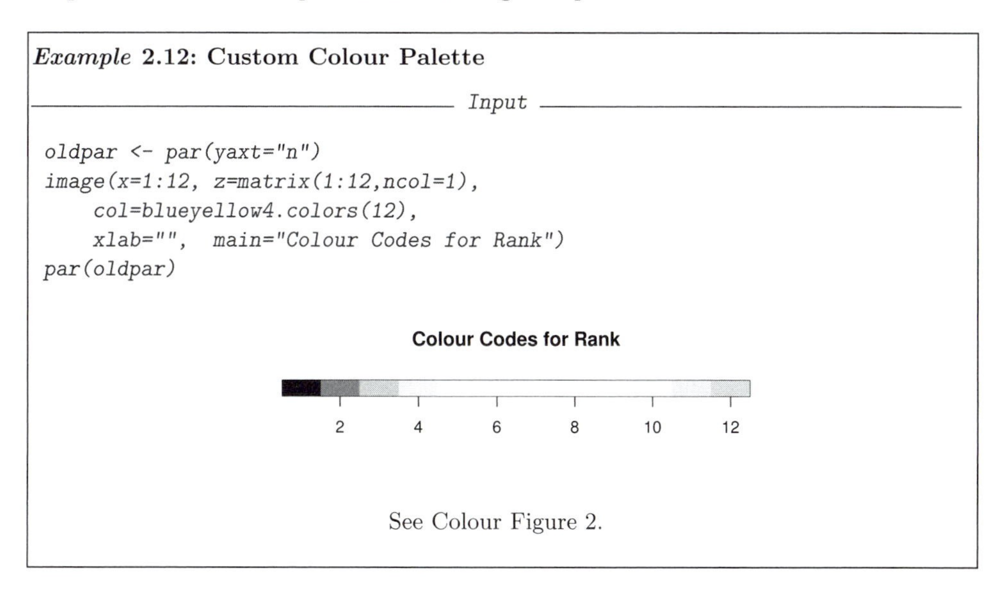

See Colour Figure 2.

To get a proper orientation and scaling, we adapt the built-in function *image()*. We need two more support functions.

Input

```
parasp<-
# set aspect ratio match data matrix or given aspect ratio
# usage: opar<-par(no.readonly=TRUE); on.exit(par(opar)); parasp(dat)
function(dat, aspr=dim(dat)[1]/dim(dat)[2], tol=0.01)
{        if (aspr<= 0) stop("parasp: aspr must be positive")
        pin<-par("pin")
        ar <- pin[1]/pin[2]
        if (abs(ar/aspr)>tol) {
                if (ar < aspr)
                        pin[2] <- pin[1]/aspr  else pin[1] <- pin[2]* aspr
                par(pin=pin)
        }
}
```

Input

```
imagem <-
# a variant of image.default keeping matrix orientation
function (z, zlim = range(z[is.finite(z)]), xlim = c(1,ncol(z)),
    ylim = c(1,nrow(z)), col = heat.colors(12),
    add = FALSE, xaxs = "i", yaxs = "i", xlab, ylab,main,
    breaks, oldstyle=FALSE,
    names=TRUE, coloffs=-1, rowoffs=4,...)
{ textnames <-
    function (zi, coloffs=-1, rowoffs=NULL) {
    # note: image interchanges rows/colums
    for (x in (1:dim(zi)[1]) ) # column labels
        text(x, ncol(zi)+0.5, rownames(zi)[x], pos=3,
        xpd=NA, offs= coloffs, srt=270)

    r <- par("usr")[2]
    for (y in (1:dim(zi)[2]))  # row labels
        text(r, y, colnames(zi)[y], pos=4, xpd=NA, offs=rowoffs,srt=0)
    } # textnames

    zi <- t(z)
    opin <- par("pin"); on.exit(par(pin=opin))
    parasp(zi)
    image(
    1:nrow(zi),1:ncol(zi), zlim=zlim,
    #xlim=xlim,
    ylim=c(ncol(zi)+0.5,0.5),
    col=col, add=add, xaxs=xaxs, yaxs=yaxs,
    xlab="", ylab="",z=zi,
    main=main,
    breaks=breaks, oldstyle=oldstyle,
#                frame.plot=FALSE,
    ...)
    if (names) {
        textnames(zi,coloffs=-4,rowoffs=1)
    } # names
}#imagem
```

Now we apply these functions to our data:

***Example* 2.13: Residual Diagnostic**

Input

```
a35 <- as.matrix( p35[3:10] )
a35rk <- apply(a35, 2, rank)
imagem(t(a35rk), col=blueyellow4.colors(10), main="p35")
```

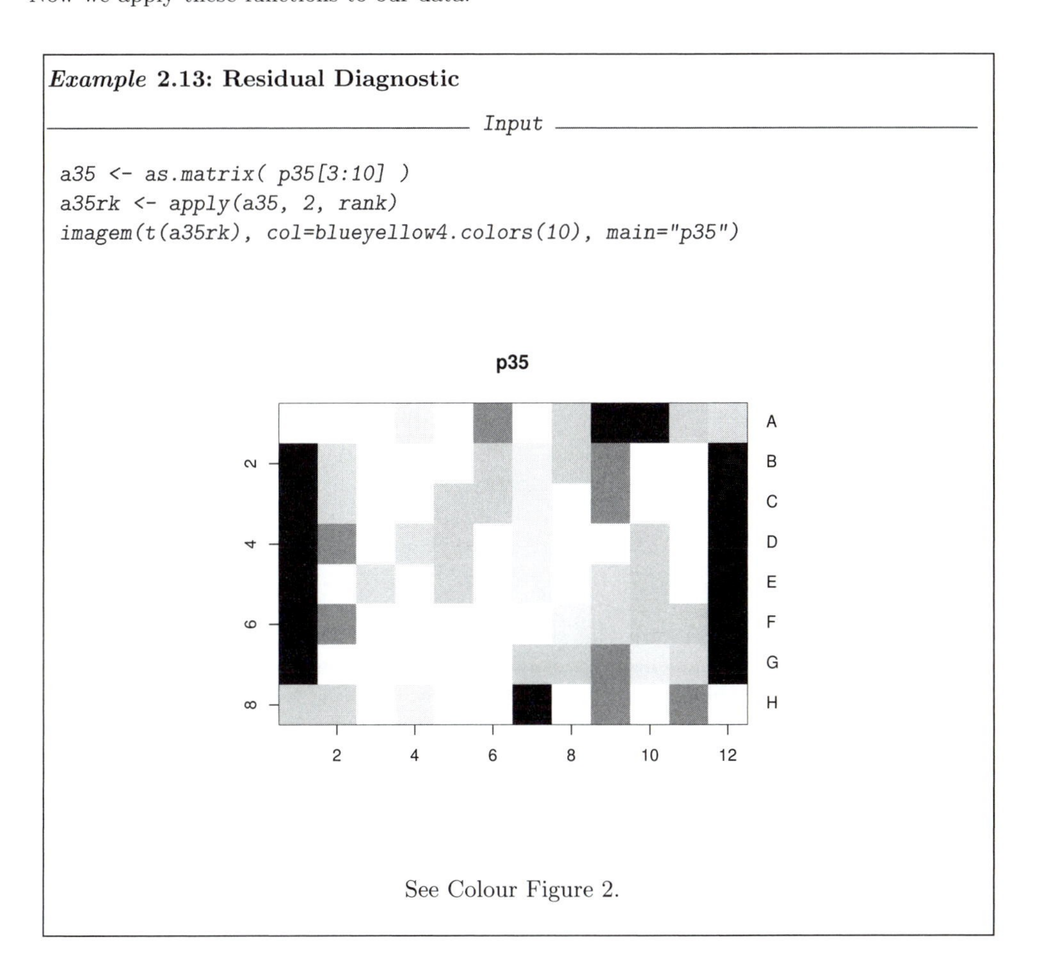

See Colour Figure 2.

For independent errors we would expect a random distribution of the rank values over the rows. The concentration of the extreme values in the extreme columns shows an inhomogeneity in the experimental production process.

So, for this example, we can report that apparently there is a difference between treatment A and certain other treatments. This finding has to be viewed cautiously: the model assumptions are not satisfied. There is a noticeable inhomogeneity between the lanes (columns). More important than the formal findings on the treatment differences may be the hint to look into the production process for the sources of this inhomogeneity.

We would not be able to detect a similar effect for the rows. Since the treatments are applied row-wise, treatment effect and row effect are confounded, and the effects are not identifiable.

2.5 Beyond Linear Regression

Transformations

Finding scale and location parameters can be understood as an attempt to transform the distribution to some reference form, however, scale and location cover only linear transformations.

The ***Box-Cox transformations***

$$y^{(\lambda)} = \begin{cases} \frac{y^{\lambda}-1}{\lambda} & \text{for} \lambda \neq 0, \\ \log(y) & \text{for} \lambda = 0 \end{cases}$$

are a family that is scaled to imbed the logarithm transformation smoothly in the power transformations. The function `boxcox()` in `library(MASS)` can be used to choose λ.

Using transformations, linear models can sometimes be used to cover non-linear situations.

Generalised linear models are an extension of linear models that allow us to cover certain transformations within the model. For more information, see [52].

2.5.1 Generalised Linear Models

We want to proceed to practical work. But at this point we should consider how to overcome the limiting assumptions of linear models. Linear models are among the best-investigated statistical models. Theory and algorithms are far advanced. So it is tempting to try to extend this class of models while still allowing ourselves to use the theoretical and algorithmic know-how.

We have formulated the linear model as

$$\begin{aligned} Y &= m(X) + \varepsilon \\ & Y \text{ with values in } \mathbb{R}^n \\ & X \in \mathbb{R}^{n \times p} \\ & E(\varepsilon) = 0 \\ & \text{with } m(X) = X\beta, \quad \beta \in \mathbb{R}^p. \end{aligned}$$

An important extension is to remove the linearity assumption. As an intermediate step, we do not suppose any longer that m is linear, but only that it can be factored using a linear function. This results in a generalised linear model

$$\begin{aligned} Y &= m(X) + \varepsilon \\ & Y \text{ with values in } \mathbb{R}^n \\ & X \in \mathbb{R}^{n \times p} \\ & E(\varepsilon) = 0 \\ & m(X) = \overline{m}(\eta) \text{ with } \eta = X\beta, \ \beta \in \mathbb{R}^p. \end{aligned}$$

The next generalisation at hand is to allow for a transformation for Y. Many more generalisations have been discussed. A small number of them have proven tractable. Most important among these is a group of models called generalised linear models (GLM). Generalised linear models have extensive support in R. For most of the functions in R for linear models, there is a corresponding function for generalised linear models. For more information see `help(glm)`.

2.5.2 Local Regression

We now make a big jump. We have discussed linear models. We know that we can represent non-linear functions using linear models, but the term entering the functions must be specified independent of the data. Too many terms lead to over-fitting. So it is not the best idea to leave it to the algorithm for linear models to do the model selection. Statistical treatment of a regression problem with mild assumptions on the model function remains a problem.

A partial approach for a solution comes from analysis. In analysis, it is a standard technique to use local approximations for functions. The analogous approach is to use a localised variant instead of a global estimation. We still assume that

$$\begin{aligned} Y &= m(X) + \varepsilon \quad Y \in \mathbb{R}^n \\ & X \in \mathbb{R}^{n \times p} \\ & E(\varepsilon) = 0, \end{aligned}$$

but assume linearity only locally:

$$m(x) \approx x'\beta_{x_0} \quad \beta_{x_0} \in \mathbb{R}^p \text{ and } x \approx x_0.$$

For practical work, abstract asymptotics is not sufficient. We have to specify what $\approx$ means. This can be done with reference to some scales (for example, $x \approx x_0$ if $|x - x_0| < 3$) or with reference to the design (for example, $x \approx x_0$ if $\#i : |x - x_i| \leq |x - x_0| < n/3$). Up-to-date implementations use refinements that go beyond the scope of this introduction. For an illustration of concepts, consider the following:

Localised Gauss-Markov estimator:

For $x \in \mathbb{R}^p$, find

$$\delta = \min_d : (\#i : |x - x_i| \leq d) \geq n \cdot f,$$

where f is a chosen fraction (for example, 0.5).

Calculate the Gauss-Markov estimator $\widehat{\beta}_x$, using only those observations for which $|x - x_i| \leq \delta$. Estimate

$$\widehat{m}(x) = x'\widehat{\beta}_x.$$

This coarsening ignores all measurement points that have a distance of more than δ. More subtle methods use a weight function to reduce the influence of distant measurement points progressively.

---

help(loess)

`loess`	*Local Polynomial Regression Fitting*

Description

Fit a polynomial surface determined by one or more numerical predictors, using local fitting.

Usage

```
loess(formula, data, weights, subset, na.action, model = FALSE,
      span = 0.75, enp.target, degree = 2,
      parametric = FALSE, drop.square = FALSE, normalize = TRUE,
      family = c("gaussian", "symmetric"),
      method = c("loess", "model.frame"),
      control = loess.control(...), ...)
```

Arguments

`formula`	a formula specifying the numeric response and one to four numeric predictors (best specified via an interaction, but can also be specified additively). Will be coerced to a formula if necessary.
`data`	an optional data frame, list or environment (or object coercible by `as.data.frame` to a data frame) containing the variables in the model. If not found in `data`, the variables are taken from `environment(formula)`, typically the environment from which `loess` is called.
`weights`	optional weights for each case.
`subset`	an optional specification of a subset of the data to be used.
`na.action`	the action to be taken with missing values in the response or predictors. The default is given by `getOption("na.action")`.
`model`	should the model frame be returned?
`span`	the parameter α which controls the degree of smoothing.
`enp.target`	an alternative way to specify `span`, as the approximate equivalent number of parameters to be used.
`degree`	the degree of the polynomials to be used, up to 2.
`parametric`	should any terms be fitted globally rather than locally? Terms can be specified by name, number or as a logical vector of the same length as the number of predictors.
`drop.square`	for fits with more than one predictor and `degree=2`, should the quadratic term (and cross-terms) be dropped for particular predictors? Terms are specified in the same way as for `parametric`.
`normalize`	should the predictors be normalized to a common scale if there is more than one? The normalization used is to set the 10% trimmed standard deviation to one. Set to false for spatial coordinate predictors and others know to be a common scale.
`family`	if `"gaussian"` fitting is by least-squares, and if `"symmetric"` a re-descending M estimator is used with Tukey's biweight function.
`method`	fit the model or just extract the model frame.
`control`	control parameters: see `loess.control`.
`...`	control parameters can also be supplied directly.

Details

Fitting is done locally. That is, for the fit at point x, the fit is made using points in a neighbourhood of x, weighted by their distance from x (with differences in 'parametric' variables being ignored when computing the distance). The size of the neighbourhood is controlled by α (set by `span` or `enp.target`). For $\alpha < 1$, the neighbourhood includes proportion α of the points, and these have tricubic weighting (proportional to $(1 - (\text{dist/maxdist})^3)^3$. For $\alpha > 1$, all points are used, with the 'maximum distance' assumed to be $\alpha^{1/p}$ times the actual maximum distance for p explanatory variables.

For the default family, fitting is by (weighted) least squares. For `family="symmetric"` a few iterations of an M-estimation procedure with Tukey's biweight are used. Be aware that as the initial value is the least-squares fit, this need not be a very resistant fit.

It can be important to tune the control list to achieve acceptable speed. See `loess.control` for details.

Value

An object of class `"loess"`.

Note

As this is based on the `cloess` package available at `netlib`, it is similar to but not identical to the `loess` function of S. In particular, conditioning is not implemented.

The memory usage of this implementation of `loess` is roughly quadratic in the number of points, with 1000 points taking about 10Mb.

Author(s)

B. D. Ripley, based on the `cloess` package of Cleveland, Grosse and Shyu available at `http://www.netlib.org/a/`.

References

W. S. Cleveland, E. Grosse and W. M. Shyu (1992) Local regression models. Chapter 8 of *Statistical Models in S* eds J.M. Chambers and T.J. Hastie, Wadsworth & Brooks/Cole.

See Also

`loess.control`, `predict.loess`.
`lowess`, the ancestor of `loess` (with different defaults!).

Examples

```
cars.lo <- loess(dist ~ speed, cars)
predict(cars.lo, data.frame(speed = seq(5, 30, 1)), se = TRUE)
# to allow extrapolation
cars.lo2 <- loess(dist ~ speed, cars,
  control = loess.control(surface = "direct"))
predict(cars.lo2, data.frame(speed = seq(5, 30, 1)), se = TRUE)
```

While linear regression is forced by the model assumptions to give a linear image (or an image with linear parametrisation) under all circumstances, a localised variant can represent non-linearities properly. The analysis of this family of methods constitutes a branch of its own in statistics: nonparametric regression.

We prepare an example:

Input

```
x <- runif(50) * pi
y <- sin(x)+rnorm(50)/10
```

***Example* 2.14: Non-Linear Regression**

Input

```
plot(x, y)
abline(lm(y ~ x), lty = 3, col = "blue")
lines(loess.smooth(x, y), lty = 6, col = "red")
legend("topleft",
    legend = c("linear", "loess"),
    lty = c(3, 6), col = c("blue", "red"), bty = "n")
```

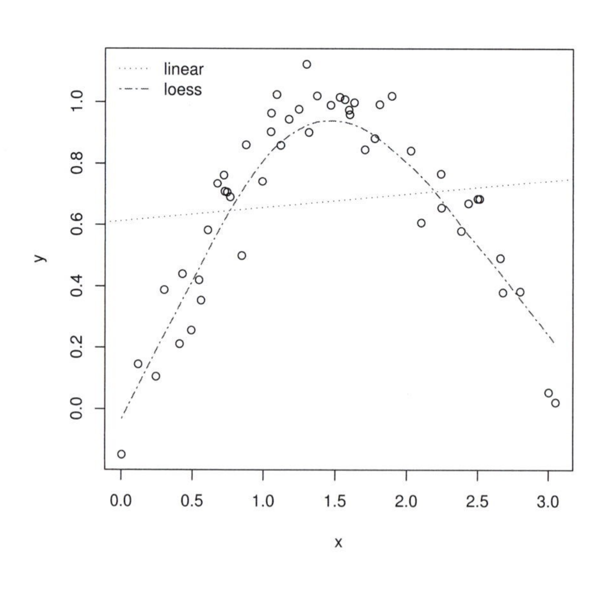

2.6 R Complements

2.6.1 Complements: Discretisation

In analogy to the construction of histograms we can discretise the data. We did this already with respect to the regressors when we discussed the helicopter example. The response can be discretised as well. This converts a regression problem into a contingency table problem. We will not delve into this possibility here.

2.6.2 Complements: External Data

Data, as well as any other R objects, can be written into an external file using `save()`, and read from there again using `load()`. Data are stored in these files in compressed form. The R storage files are exchangeable between different R systems.

Data In/Output for R	
`save()`	stores data in an external file. *Syntax:* `save(⟨names of the objects to store⟩, file = ⟨file name⟩, ...)`
`load()`	loads data from an external file. *Syntax:* `load(file = ⟨file name⟩, ...)`

Often data are collected or prepared with other systems. R provides a series of functions to read data in various formats. See Appendix A.15 (page A-215). More information is given in the manual "Data Import/Export" ([32]).

The function `data()` bundles various access routines, provided access paths and file names follow the R conventions.

Often imported data have to be post-processed to align the format to the calling conventions of the R functions that will be applied.

For example, `lm()` expects the regressors to be separate variables. On the other side, for factorial designs it is usual to collect the results in a table where the factor levels appear as row or column labels. The function `stack()` transforms tables into columns. See also Appendix A.9 (page A-205).

2.6.3 Complements: Testing Software

All algorithms, and so the algorithms for linear models and their variants, should be treated with the same precautions as any mathematical publication or quotation. Unfortunately, even simple programs soon have a semantic complexity far beyond that of many mathematical proofs. The usual strategies like re-computing or stepwise execution forbid themselves. Selective testing must take the place of complete examination. An example of how to design test strategies is documented in [42].

Scrutiny is necessary for the implementation as well as for the underlying abstract algorithm.

Exercise 2.15	
	For this series of exercises, let $y_i = a + bx_i + \varepsilon_i$ with ε_i iid $\sim N(0, \sigma^2)$ and $x_i = i, i = 1, \ldots, 10$.
	Choose a strategy to inspect `lm()` with regard to the parameter space (a, b, σ^2). Are there apparent cellular decompositions for the parameters a, b, σ^2? What are the trivial cases? What are the asymptotics that apply? Choose test points in the interior of each cell and on the boundaries. Perform these tests and summarise the results.
	What are the symmetries/anti-symmetries that apply? Check for these symmetries.
	Which invariant or covariate behaviour applies? Check for these invariant or covariate behaviour.

Exercise 2.16	
	For this series of exercises, let $y_i = a + bx_i + \varepsilon_i$ with ε_i iid $\sim N(0, \sigma^2)$.
	What are the extremal designs (x_i)? Check the behaviour of `lm()` for four extremal designs.
	Perform the tests from the last exercise, now with variable design. Summarise your results.

Exercise 2.17	
	For this series of exercises, let $y_i = a + bx_i + \varepsilon_i$ with ε_i iid $\sim N(0, \sigma^2)$.
	Modify `lm()` to give a fail-safe function for simple linear models that checks deviations from the model assumptions as well.

2.6.4 R *Data Types*

R is an interpreted programming language. It tries to allow a flexible handling of definitions and specifications for the user. For performance reasons, R attempts to evaluate expressions and terms as late as possible. This requires some restrictions for the language, which makes R differ from other programming languages.

R does not have abstract data types. A data type is defined by its instances, the variables.

The data type of a variable is dynamic: in the same context the same name can refer to different variable values and different variable types at different times.

At any time, however, any variable has a well-defined type. The R type system is best understood with reference to its historical development and the corresponding functions. Initially, the type was described by `mode()` (for example, "numeric") and `storage.mode()` (for example, "integer" or "real").

Essentially both functions have been replaced by `typeof()`. A summary of the information reported by `typeof()` is in [33].

More complex data types are derived from the basic types documented in [33] by attaching attributes to the variables. This is done using function `attr()`, which can also be applied to inspect attributes. So, for example, a matrix or an array are just special vectors, which are special by having a `dim` attribute. The `class` attribute serves to denote and name a class explicitly.

Type test and conversion functions are supplied for the main types: `is.`⟨type⟩`()` tests for a type, `as.`⟨type⟩`()` converts to a type.

For the exceptional values, there are special test functions `is.inf()`, `is.na()` and `is.nan()`. A `NaN` value is always considered as missing, so both `is.na(NaN)` and `is.nan(NaN)` will return `TRUE`.

See also Appendix A.4 (page A-197).

2.6.5 Classes and Polymorphic Functions

As the system developed further, concepts from object-oriented programming have been taken into R. The special attribute `class` is used here. The name of the type (or the "class") is stored as the `class` attribute. Multiple class membership is possible. In this case, `class` is a vector of class names. So, for example, an ordered factor has the attribute `class = c("ordered", "factor")`. For class management, the functions `class()`, `unclass()`, `inherits()` are provided.

Class membership in R is based on trust. R does not check whether a data structure is consistent with its acclaimed class.

Functions like `plot()`, `print()` and many others inspect the class or type of its argument and branch to specialised functions as appropriate. This is called function polymorphism. If you list a polymorphic function, first you get only a clue that apart from special cases a dispatch function `UseMethod()` is called. For example:

Example 2.15: Method Dispatch

Input

```
plot
```

Output

```
function (x, y, ...)
{
    if (is.function(x) && is.null(attr(x, "class"))) {
        if (missing(y))
            y <- NULL
        hasylab <- function(...) !all(is.na(pmatch(names(list(...)),
            "ylab")))
        if (hasylab(...))
            plot.function(x, y, ...)
        else plot.function(x, y, ylab = paste(deparse(substitute(x)),
            "(x)"), ...)
    }
    else UseMethod("plot")
}
<environment: namespace:graphics>
```

UseMethod() tries to identify the class of the first argument that was used in the function call. Then it tries to find a specialisation of the function, a method, for this class and finally it calls this function. For ***polymorphic*** functions the available methods are listed with *methods()*, for example, *methods(plot)*.

2.6.6 Extractor Functions

Functions like *lm()* yield complex data with rich information. In a purely object-oriented environment, access methods and data would be encapsulated. In R, object orientation is implemented rudimentarily and with varying models. This partially reflects the historical development of the language. For sufficiently general structures, access methods as in Section 2.6.5 (page 103) are available. For the objects as returned by *lm()*, for example, a series of extractor functions provide access to components in various forms.

Extractor Functions for `lm`	
coef()	extracts the estimated coefficients.
effects()	extracts successive orthogonal components.
residuals()	raw residuals.
stdres()	(in *library(MASS)*) standardised residuals.

(cont.)→

Extractor Functions for `lm` (cont.)	
studres()	(in *library(MASS)*) externally standardised residuals.
fitted()	extracts fitted values from objects returned by modeling functions.
vcov()	variance/covariance matrix of the estimated parameter.
predict()	confidence and tolerance intervals.
confint()	confidence intervals for parameter.
influence()	extracts influence diagnostics.
model.matrix()	derives the design matrix.

2.7 Statistical Summary

The leading example in this chapter was the statistical analysis of a functional relationship. The models considered are finite in the sense that only a finite-dimensional function space was taken into account to describe the relation between regressors and response. The stochastic component in these models was restricted to a (one-dimensional) random variable. The notion of dimension here deserves further attention. For one, we have the regressor dimension. This is the dimension of the space of the observed or derived parameters. Not all parameters are identifiable or estimable. More precisely, this dimension is the vector space dimension of the chosen model space. In matrix terms, this is the dimension of the space spanned by the columns of the design matrix. Models are indicated by parameters in this space. The parameters can be unknown or hypothetical. In any case, we have considered the parameters as deterministic. On the other hand we have the stochastic component, represented by the error term. In this chapter, we assumed homogeneous errors. So the error term in principle is one-dimensional, with distribution taken from a given space of distributions. For the special case of simple Gauss-linear models, the distributions are specified by two parameters, the expected value and the variance. We made the assumption that the model would cover all systematic effects on average, hence the expected value of the error is zero. The variance remains an unknown nuisance parameter. We avoided the problems resulting from this nuisance parameter by restricting ourselves to problems where the nuisance parameter could be replaced by an estimated value and thus be eliminated.

2.8 Literature and Additional References

[5] Chambers, J.M.; Hastie, T.J. (eds.) (1992): *Statistical Models in S.* Chapman & Hall, New York.

[22] Jørgensen, B. (1993): *The Theory of Linear Models.* Chapman & Hall, New York.

[33] R Development Core Team (2004–2008): The R language definition.

[42] Sawitzki, G. (1994): Numerical Reliability of Data Analysis Systems. *Computational Statistics & Data Analysis* 18.2 (1994), 269–286. `<http://www.statlab.uni-heidelberg.de/reports/>`.

[43] Sawitzki, G. (1994):Report on the Numerical Reliability of Data Analysis Systems. *Computational Statistics & Data Analysis/SSN* 18.2 (1994) 289–301. `<http://www.statlab.uni-heidelberg.de/reports/>`.

CHAPTER 3

Comparisons

We begin with the construction of a small gadget that will provide us with example data. The base is a reaction tester. We present a "random" point, wait for a mouse click on that point and record the position of the mouse pointer. To get a stable image for repeated activations, we fix the coordinate system.

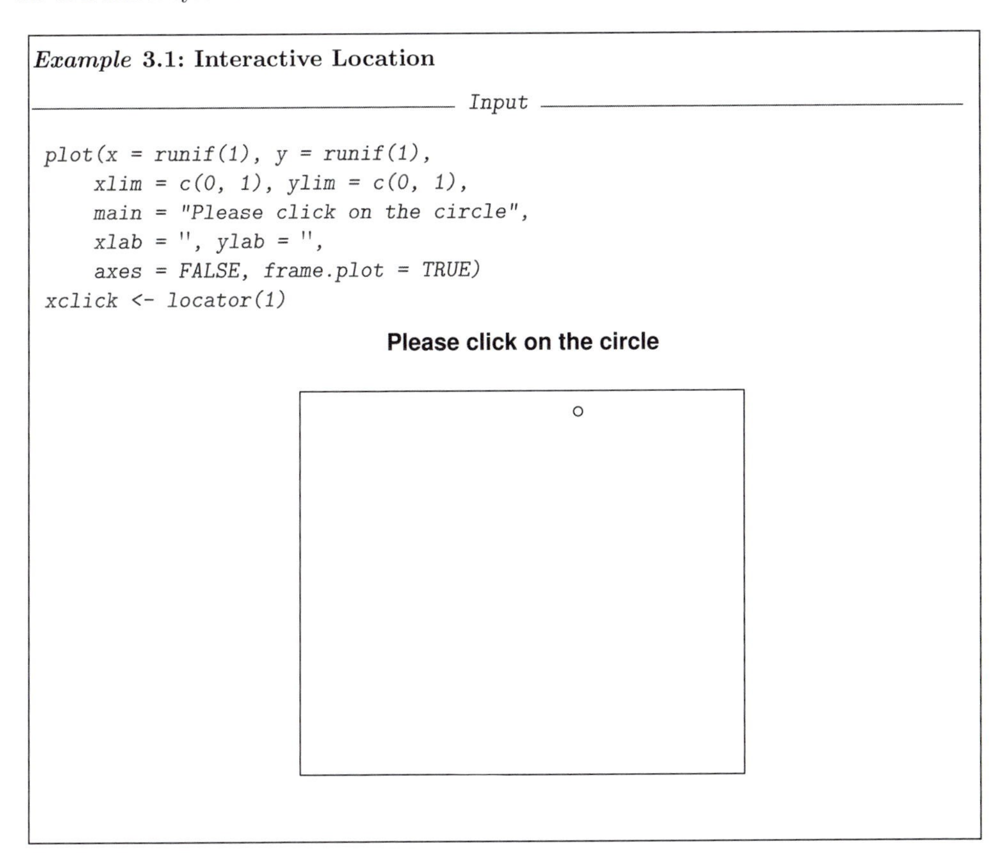

***Example* 3.1: Interactive Location**

Input

```
plot(x = runif(1), y = runif(1),
    xlim = c(0, 1), ylim = c(0, 1),
    main = "Please click on the circle",
    xlab = '', ylab = '',
    axes = FALSE, frame.plot = TRUE)
xclick <- locator(1)
```


Now we wrap up the base function in a timer. We record the coordinates, try to measure the reaction time, and return the results as a list.

Example 3.2: Click Timing

Input

```
click1 <- function(){
    x <- runif(1);y <- runif(1)
    plot(x = x, y = y, xlim = c(0, 1), ylim = c(0, 1),
        main = "Please click on the circle",
        xlab = '', ylab = '',
        axes = FALSE, frame.plot = TRUE)
    clicktime <- system.time(xyclick <- locator(1))
    list(timestamp = Sys.time(),
        x = x, y = y,
        xclick = xyclick$x, yclick = xyclick$y,
        tclick = clicktime[3])
}
```

For later processing we can integrate the list in a `data.frame` and extend this `data.frame` stepwise using `rbind`.

Example 3.3: Sequential Recording

Input

```
dx <- as.data.frame(click1())
dx <- rbind(dx, data.frame(click1()))
dx
```

Output

```
                   timestamp      x      y xclick yclick tclick
elapsed  2008-12-12 15:14:26 0.7034 0.6072 0.7044 0.6070  1.899
elapsed1 2008-12-12 15:14:28 0.6948 0.9285 0.7003 0.9352  1.855
```

Exercise 3.1	Click Timing
	Define a function `click(runs)` that repeats `click1()` a chosen number *runs* of times and returns the result as a `data.frame`. An additional first timing should be considered as a "warming up" and is not included in the following evaluation. Select a number `runs`. Give reasons for your choice of `runs`. Execute `click(runs)` and store the result in a file using `write.table()`. Display the distribution of the component `tclick` with the methods from Chapter 1 (distribution function, histogram, box-and-whisker plot).

3.1 Shift/Scale Families, and Stochastic Order

A comparison of distributions can be a demanding task. The mathematical space of distributions is not a number space any more or a finite-dimensional vector space. Properly, the room in which distributions reside is a space of probability measures. In simple cases, such as for distributions on $\mathbb{R}$, we can reduce everything to distribution functions and come to a tractable function space. However, even here a comparison can cause big problems. We do not have any simple order relation upon which to base the comparison.

Exercise 3.2	Click Comparison
	Perform Exercise 3.1 using the right hand and then again using the left hand. Compare the empirical distributions of the timing data returned by `tclick()`.
	The recorded data also contain information about the positions. Define a distance measure `dist` for the deviation. Give reasons for your definition. Perform a right/left comparison for `dist`.

We concentrate on the comparison of two distributions only, for example, that of the results of two treatments. And we take a simple case: we assume that the observations are independent and identically distributed for each treatment. We use the index notation that is usual for the comparison of treatments in the two sample case.

Y_{ij} independent identically distributed with distribution function F_i

$i = 1, 2$ treatments

$j = 1, \ldots, n_i$ observations in treatment group i.

How do we compare the observations in the treatment groups $i = 1, 2$? The (simple) linear models

$$Y_{ij} = \mu + \alpha_i + \varepsilon_{ij}$$

consider only the case where the difference amounts to a shift $\Delta = \alpha_1 - \alpha_2$.

Notation: For a distribution with distribution function F the family

$$F_a(x) = F(x - a)$$

is called the ***shift family*** for F. The parameter a is called the shift or location parameter.

Speaking in terms of probabilities, the treatment can shift probability mass in quite different ways from what can be achieved by an additive shift term. We need more general ways to compare distributions. Shift families are not the only framework to consider.

Notation: A distribution with distribution function F_1 is ***stochastically smaller*** than a distribution with distribution function F_2 (in symbols, $F_1 \prec F_2$), if a variable distributed as F_1 takes rather smaller values than a variable distributed as F_2. This means that F_1 increases sooner.

$$F_1(x) \geq F_2(x) \ \forall x$$

and

$$F_1(x) > F_2(x) \text{ for at least one } x.$$

For shift families we have: If $a < 0$, then $F_a \prec F$. The shift results in a parallel shift of the distribution functions.

A typical result of the click comparison experiment (Exercise 3.2) is given in Figure 3.1. The response times for the right side are stochastically smaller than those for the left side. But the distributions do not belong to a common shift family, since the distribution functions are not parallel.

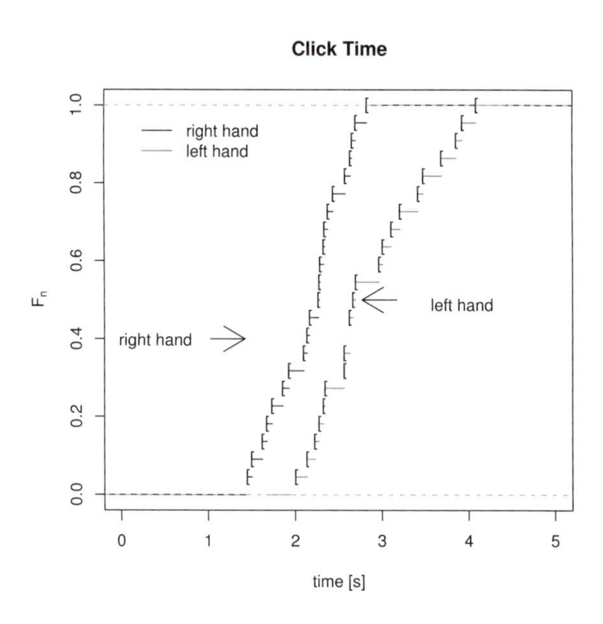

Figure 3.1 *Distribution functions for the right/left click time*

Exercise 3.3	**Stochastic Order**
	What does a PP plot for F_1 against F_2 look like if $F_1 \prec F_2$?
	What does a QQ plot for F_1 against F_2 look like if $F_1 \prec F_2$?

Unfortunately, the stochastic order defined here is only of limited value. It does not define a complete order. For shift families it is sufficient. But counter examples can be constructed by minor extensions of shift families.

Notation: For a distribution with distribution function F the family with

$$F_{a,b}(x) = F(\frac{x-a}{b})$$

is called the ***scale shift family*** for F.

Exercise 3.4	
	The scale shift family for the $N(0,1)$ distribution are the $N(\mu,\sigma^2)$ distributions. Which $N(\mu,\sigma^2)$ distributions are stochastically smaller than the $N(0,1)$ distribution? Which are stochastically larger? Which distributions have an undefined order relation to $N(0,1)$?

The ordering by scale and location, which comes from the linear theory, and stochastic order are different concepts. Both aspects must be considered separately. Many statistical methods concentrate on aspects that are motivated by scale shift families. Differences going beyond what can be formulated in terms of scale and shift often need special attention.

In Chapter 2 we considered a typical situation for linear models. In principle we dealt with scale shift families. The scale parameter in these models is but a nuisance parameter that can be eliminated. For this, we used an estimator for this stochastic scale parameter, the residual variance, which we cancelled out ultimately. As a peculiarity of Gauss-linear models we get independent estimators for the expected value and the variance. This helps us in Gauss-linear models to derive statistics that are independent of the nuisance scale parameter. The remainder of Chapter 2 only considered shift alternatives.

In general, however, we have a spectrum of problems:

- shift alternatives
- shift/scale alternatives
- stochastic ordering
- general alternatives

Testing and estimation methods often concentrate on just one aspect of the problem, the location. The scale parameter of the stochastic error here is only a nuisance parameter. Differences in the shift parameter lead to stochastically monotone relations. Differences in the scale parameter are not categorised so easily. If both the shift parameter and the scale parameter vary, the situation is more complicated and test statistics must first be cleared from the scale as a nuisance parameter.

3.2 *QQ* Plot, *PP* Plot, and Comparison of Distributions

As means to compare distributions we already encountered the *PP* plot and the *QQ* plot in Chapter 1. In Chapter 1 we used these plots to compare an empirical distribution with a theoretical distribution, but, of course, with some caution, we can apply it to compare two empirical distributions.

As long as we stay within one shift scale family, the *QQ* plot has at least one advantage over the *PP* plot:

Remark 3.1 *If F_1, F_2 are distribution functions from a common shift scale family, the QQ plot of F_1 against F_2 is a straight line.*

In particular for the Gaussian distributions the *QQ* plot against $N(0,1)$ is a valuable tool. Any normal distribution gives a straight line in this plot. The *QQ* plot is already implemented for this situation as function `qqnorm()`.

For the comparison of two samples the corresponding function *qqplot()* can be used: if we denote the empirical quantiles by $Y_{1,(i:n)}$ resp. $Y_{2,(i:n)}$, the plot is the graph $\left(Y_{1,(i:n)}, Y_{2,(i:n)}\right)_{i=1\ldots n}$ in the case of equal sample sizes. The necessary caution we mentioned above comes in if the sample sizes differ. If the sample sizes differ, R generates nodes by linear interpolation, with the smaller of the two sample sizes determining the number of nodes.

The *PP* plot does not have equivariance properties comparable to those of the *QQ*. If we want to eliminate scale and shift parameters, we have to transform the data first. Mathematical theory, however, is simple for the *PP* plot. In particular, there is a corresponding Kolmogorov-Smirnov test (see Section 3.2.1 (page 116)).

On the other hand, the equivariance properties of the *QQ* plots are paid for with structural deficits. In low density regions, few data points control the plot. As a consequence, the plot shows high variance. But in these regions we usually are in the tails of a distribution and quantiles that are near in probability are far apart in value space: high variance combines unfavourably with high variability, and correspondingly the *QQ* plot shows large fluctuation. For most textbook distributions this means that the *QQ* plot is not very useful in the tails. The *PP* plot does not have this kind of scale deficit, but then it does not have the equivariance properties of the *QQ* plots. To cope with this, the *PP* plot is generally applied to variables that are standardised in an appropriate way. We will give an example soon.

---

help(qqplot)

---

qqnorm	*Quantile-Quantile Plots*

Description

qqnorm is a generic function the default method of which produces a normal QQ plot of the values in y. qqline adds a line to a normal quantile-quantile plot which passes through the first and third quartiles.

qqplot produces a QQ plot of two datasets.

Graphical parameters may be given as arguments to qqnorm, qqplot and qqline.

Usage

```
qqnorm(y, ...)
## Default S3 method:
qqnorm(y, ylim, main = "Normal Q-Q Plot",
       xlab = "Theoretical Quantiles", ylab = "Sample Quantiles",
       plot.it = TRUE, datax = FALSE, ...)

qqline(y, datax = FALSE, ...)

qqplot(x, y, plot.it = TRUE, xlab = deparse(substitute(x)),
       ylab = deparse(substitute(y)), ...)
```

Arguments

`x`	The first sample for `qqplot`.
`y`	The second or only data sample.
`xlab, ylab, main`	plot labels. The `xlab` and `ylab` refer to the y and x axes respectively if `datax = TRUE`.
`plot.it`	logical. Should the result be plotted?
`datax`	logical. Should data values be on the x-axis?
`ylim, ...`	graphical parameters.

Value

For `qqnorm` and `qqplot`, a list with components

`x`	The x coordinates of the points that were/would be plotted
`y`	The original `y` vector, i.e., the corresponding y coordinates *including NAs*.

References

Becker, R. A., Chambers, J. M. and Wilks, A. R. (1988) *The New S Language*. Wadsworth & Brooks/Cole.

See Also

`ppoints`, used by `qqnorm` to generate approximations to expected order statistics for a normal distribution.

Examples

```
require(graphics)

y <- rt(200, df = 5)
qqnorm(y); qqline(y, col = 2)
qqplot(y, rt(300, df = 5))

qqnorm(precip, ylab = "Precipitation [in/yr] for 70 US cities")
```

Exercise 3.5	
	Use the *QQ* plot to compare the results of the right/left `click` experiments. Summarise the results.
	(cont.)→

Exercise 3.5	(cont.)
	Combine the right/left `tclick` data to a vector. Compare the *QQ* plot with that of Monte Carlo samples taken from the joined vector. *Hint:* You can draw random samples with `sample()`. With `par(mfrow = c(2, 2))` you arrange the display area so that it shows four plots at a time.
**	For `sample()` use `replace = FALSE`. How do you have to apply `sample()` now to split the joint vector into two vectors with Monte Carlo samples? What differences do you expect in comparison to `replace = TRUE`?

Exercise 3.6	
	Find scale and shift parameters for the right/left `click` data so that, after using these parameters for transformation, the groups match as well as possible. Describe the differences using these parameters. Use a model formulation in terms of a linear model.
	Use the function `boxplot()` to display quartiles and tail behaviour. Compare the information with the information you derived from the scale and shift parameters. *Hint:* What corresponds to the shift (or location) parameter? What corresponds to the scale parameter?

If representations such as visual representations in displays or numeric representations in summary statistics are affine invariant, scale and shift parameters can be ignored. If representations are not affine invariant, it is often helpful to estimate scale and shift parameters first, then standardise the distributions, and only then to inspect the standardised distributions.

The potential problem with this is that we have to take into account the stochastic behaviour of the scale and shift parameter estimation. The usual way out is to be cautious and use "conservative" tests and robust estimators. The following function tries to transform scale and location to match a standard normal distribution.

```
ScaleShiftStd <- function (x) {
   xq <- quantile(x[!is.na(x)], c(0.25, 0.75))
   y <- qnorm(c(0.25, 0.75))
   slope <- diff(y)/diff(xq)
   (x-median(x, na.rm = FALSE)) * slope
   }
```

Exercise 3.7	**Scale/Shift Standardisation**
	This algorithm is only appropriate for symmetric distributions. Combine it with a power transformation as in Section 2.5 (page 96) to symmetrise a distribution and give an algorithm that can be applied to asymmetric transformations.

For direct comparison of distributions, we return to techniques discussed in the first chapter. In principle what was said there about the comparison to a theoretical distribution can be carried over to the comparison of two distributions, for example, to the comparison of two treatments. The statistical statements, however, have to be adapted. In Chapter 1 we had one fixed hypothetically known distribution to be compared with an empirical distribution. This is sometimes called the ***one-sample case***. Now we are in the ***two-sample case***. We have to compare two empirical distributions, with the understanding that the proper targets of comparison are the underlying unknown distributions.

The Monte Carlo band that we used in the one-sample case does not have an immediate counterpart. We want to compare two distributions, but we have no distinguished model distribution from which to draw reference samples.

We can, however, modify the idea and construct conditional Monte Carlo bands. Conditioning here means the construction is dependent on the observed sample values. Assume we have two samples $Y_{11}, \ldots, Y_{1n_1}$ and $Y_{21}, \ldots, Y_{2n_2}$ of independent and, within the group, identically F_1 resp. F_2 distributed observations. If there is no difference between the groups, $(Y_{11}, \ldots, Y_{1n_1}, Y_{21}, \ldots, Y_{2n_2})$ is an iid random sample from the common distribution $F = F_1 = F_2$ with sample size $n = n_1 + n_2$. For an iid random sample, each permutation of the indices would arise with equal probability.

This motivates the following procedure: we permute the tuple $(Y_{11}, \ldots, Y_{1n_1}, Y_{21}, \ldots, Y_{2n_2})$ and after permutation we assign the first n_1 values to the first group, the remaining values to the second.

As n_1, n_2 increase, the permutation group soon is so big that an exhaustive evaluation becomes impossible. As a way out we use a random sample of permutations. We use the values generated by this procedure to generate Monte Carlo bands.

Exercise 3.8	Two-Sample Monte Carlo Bands
	Modify the functions for the *PP* plot and the *QQ* plot so that Monte Carlo bands for the comparison of two samples are added. (Use a scale/shift standardisation for the *PP* plot.) *Hint:* Use the function `sample()` to generate random permutations.

At larger sample sizes, the mere administrative overhead to generate permutations may become too great. Instead of generating permutations, drawing samples with replacement from the n values $(Y_{11}, \ldots, Y_{1n_1}, Y_{12}, \ldots, Y_{1n_2})$ may be more economical and may be used as a substitute. This approximate solution is known as ***bootstrap approximation***.[1]

Since there are only finitely many permutations for a given sample size, for small sample sizes we can enumerate all permutations. We choose bands such that a sufficiently high proportion of all curves are captured within the bands (for example, more than 95%). Permutations that only differ within the groups give the same bands. Hence we need not check all $n!$ permutations, but only the $\binom{n}{n_1}$ selections for the assignment to the groups.

[1] Caution: there are arbitrarily wild definitions of bootstrap. Always try to get a concise mathematical formulation when speaking about bootstrap.

Exercise 3.9	
**	Augment the *PP* plot and *QQ* plot for the `click` experiments by permutation bands that cover 95% of the permutations.
*	Generate new plots from the *PP* plots and *QQ* plots by adding Monte Carlo bands from permutations. Use the envelope of 19 Monte Carlo samples. *Hint:* Use function `sample()` to draw a random sample of sample size n_1 from $x = (Y_{11}, \ldots, Y_{1n_1}, Y_{12}, \ldots, Y_{1n_2})$.
	Hint: See `help(sample)`.

Exercise 3.10	
*	Try to compare the properties of permutation bands, Monte Carlo bands and bootstrap bands on the hypothesis where $F_1 = F_2$.

If not the distributions, but only single specified parameters are to be compared, an analogous strategy can be used. For example, if we focus on shift alternatives (that is F_1 and F_2 are from a shift family, $F_1(x) = F_2(x - a)$ for some a, we can take the mean (or the median) as the parameter of interest. The procedure given above can be used analogously to test the hypothesis that the distributions are not different ($a = 0$), based on the data.

Exercise 3.11	
*	Formulate the strategies given above for intervals of single test statistic (example: mean) instead of bands. *Hint:* Instead of the two mean values for both groups, can you use a single one-dimensional statistic?

3.2.1 Kolmogorov-Smirnov Tests

In Chapter 1 we introduced the Kolmogorov-Smirnov test for comparison of a random sample $(X_i)_{i=1,\ldots,n}$ and the corresponding empirical distribution function F_n with a (fixed, given) distribution F. The critical test statistic was

$$\sup |F_n - F|.$$

We can modify this test slightly to compare two empirical distributions. Instead of a model distribution F we now have a second empirical distribution G_m from observations $(Y_j)_{j=1,\ldots,m}$ based on some underlying (unknown) distribution G. The critical test statistic is now

$$\sup |F_n - G_m|.$$

The test based on this statistic is in the two-sample Kolmogorov-Smirnov test, which is presented in many textbooks. Together with the one-sample version, it is implemented in function `ks.test()`. Translated into a graphical version, this test corresponds to the *PP* plot and allows the construction of bands for the *PP* plot.

We also can use simulation to determine bands. In contrast to the one-sample case we do not have a given distribution from which to simulate. Under the hypothesis that the distributions F and G do not differ for independent observations the joined vector $(X_1, \ldots, X_n, Y_1, \ldots, Y_m)$ is the vector of $n+m$ independent random numbers with identical distribution $F = G$. Given a data set, this relation can be used for simulation. Using a permutation π of the indices from the vector $Z = (X_1, \ldots, X_n, Y_1, \ldots, Y_m)$ a new vector Z' with $Z'_i = Z_{\pi(i)}$ is generated. The first n components are used as simulated values $(X'_i)_{i=1,\ldots,n}$, the remaining m components as simulated values $(Y'_j)_{j=1,\ldots,m}$.

Exercise 3.12	
*	Implement this algorithm and enhance the *PP* plot by adding simulated *PP* plots generated by a small number (19?) of permutations.
	Determine the permutation distribution of $\sup \lvert F_n - G_m \rvert$ from the simulation and calculate this statistic for the original data. Can you use this comparison to define a test procedure?
	The Kolmogorov-Smirnov test as implemented uses an approximation for the two-sample case. In our simulation we know that we simulate under the hypothesis. So any rejection we get is a false rejection, i.e., an error. Inspect the distribution of the error level under the simulated conditions.

Exercise 3.13	
	Use the *QQ* plot for a pair-wise comparison of the results of the helicopter experiment from Chapter 2. Summarise your results.

Exercise 3.14	
	Inspect the implementation of `qqnorm()`. Implement an analogous function for the *PP* plot and apply it to the helicopter data.

3.3 Tests for Shift Alternatives

If we make additional distribution assumptions, we can possibly choose better decision rules. For these, however, the distribution assumptions may be critical. The dependency on these distribution assumptions can be relaxed or avoided, if we can guarantee appropriate assumptions by our experimental procedure. The F-test introduced in Chapter 2 is an example of a distribution-based method. For the two-sample case this test can be re-expressed as a t-test, which also gives an indication of the direction of the difference. (The square of the t statistic is an F statistic.)

help(**t**.test)

t.test *Student's t-Test*

Description

Performs one and two sample t-tests on vectors of data.

Usage

```
t.test(x, ...)

## Default S3 method:
t.test(x, y = NULL,
       alternative = c("two.sided", "less", "greater"),
       mu = 0, paired = FALSE, var.equal = FALSE,
       conf.level = 0.95, ...)

## S3 method for class 'formula':
t.test(formula, data, subset, na.action, ...)
```

Arguments

`x`	a (non-empty) numeric vector of data values.
`y`	an optional (non-empty) numeric vector of data values.
`alternative`	a character string specifying the alternative hypothesis, must be one of `"two.sided"` (default), `"greater"` or `"less"`. You can specify just the initial letter.
`mu`	a number indicating the true value of the mean (or difference in means if you are performing a two sample test).
`paired`	a logical indicating whether you want a paired t-test.
`var.equal`	a logical variable indicating whether to treat the two variances as being equal. If `TRUE` then the pooled variance is used to estimate the variance otherwise the Welch (or Satterthwaite) approximation to the degrees of freedom is used.
`conf.level`	confidence level of the interval.
`formula`	a formula of the form `lhs ~ rhs` where `lhs` is a numeric variable giving the data values and `rhs` a factor with two levels giving the corresponding groups.
`data`	an optional matrix or data frame (or similar: see `model.frame`) containing the variables in the formula `formula`. By default the variables are taken from `environment(formula)`.
`subset`	an optional vector specifying a subset of observations to be used.
`na.action`	a function which indicates what should happen when the data contain `NA`s. Defaults to `getOption("na.action")`.
`...`	further arguments to be passed to or from methods.

Details

The formula interface is only applicable for the 2-sample tests.

`alternative = "greater"` is the alternative that `x` has a larger mean than `y`.

If `paired` is `TRUE` then both `x` and `y` must be specified and they must be the same length. Missing values are removed (in pairs if `paired` is `TRUE`). If `var.equal` is `TRUE` then the pooled estimate of the variance is used. By default, if `var.equal` is `FALSE` then the variance is estimated separately for both groups and the Welch modification to the degrees of freedom is used.

If the input data are effectively constant (compared to the larger of the two means) an error is generated.

Value

A list with class `"htest"` containing the following components:

`statistic`	the value of the t-statistic.
`parameter`	the degrees of freedom for the t-statistic.
`p.value`	the p-value for the test.
`conf.int`	a confidence interval for the mean appropriate to the specified alternative hypothesis.
`estimate`	the estimated mean or difference in means depending on whether it was a one-sample test or a two-sample test.
`null.value`	the specified hypothesized value of the mean or mean difference depending on whether it was a one-sample test or a two-sample test.
`alternative`	a character string describing the alternative hypothesis.
`method`	a character string indicating what type of t-test was performed.
`data.name`	a character string giving the name(s) of the data.

See Also

`prop.test`

Examples

```
require(graphics)

t.test(1:10,y=c(7:20))      # P = .00001855
t.test(1:10,y=c(7:20, 200)) # P = .1245    -- NOT significant anymore

## Classical example: Student's sleep data
plot(extra ~ group, data = sleep)
## Traditional interface
with(sleep, t.test(extra[group == 1], extra[group == 2]))
## Formula interface
t.test(extra ~ group, data = sleep)
```

The textbook derivation for the t-test supposes that we have independent identically distributed samples from normal distributions. In fact, weaker assumptions are sufficient. Writing the t-test statistic as

$$t = \frac{\widehat{\mu_1} - \widehat{\mu_2}}{\sqrt{(\widehat{Var\,(\widehat{\mu_1} - \widehat{\mu_2})})}} \tag{3.1}$$

we see that t is t distributed if $\widehat{\mu_1} - \widehat{\mu_2}$ has a normal distribution, $(\widehat{Var\,(\widehat{\mu_1} - \widehat{\mu_2})})$ has a χ^2 distribution, and both terms are independent. The central limit theorem guarantees that under mild conditions $\widehat{\mu_1} - \widehat{\mu_2}$ at least asymptotically has a normal distribution. Similarly under not too strict conditions $(Var\,(\widehat{\mu_1} - \widehat{\mu_2}))$ behaves asymptotically as χ^2 distributed. If both terms are (approximatively) independent, t is approximatively t distributed. Independence is guaranteed for the normal distribution, but may critically break down for other distributions.

Exercise 3.15	
*	Use a simulation to inspect the distribution of $\overline{Y}$, $\widehat{Var\,(Y)}$ and the t statistic for Y from a uniform distribution $U[0,1]$ with sample size $n = 1, \ldots, 10$. Compare the distributions from the simulation with the corresponding normal, χ^2 resp. t distribution.
	Use a simulation to inspect the distribution of $\overline{Y}$, $\widehat{Var\,(Y)}$ and the t statistic for Y from a mixture, consisting at 90% from an $N(0,1)$- and at 10% from an $N(0,10)$ distribution, with sample size $n = 1, \ldots, 10$. Compare the distributions from the simulation with the corresponding normal, χ^2 resp. t distribution.

The t-test has some robustness which can give it approximate validity outside its normal model context. There are, however, ways to free us completely from the normal distribution assumption. If we proceed as in the F-test resp. t-test, but use ranks instead of the original data, we get test procedures that are distribution-independent (at least as long as no ties occur). The Wilcoxon test is a distribution independent variant of the t-test. Theoretically it corresponds exactly to a t-test, applied to the (jointly) rank-transformed data. Like the t-test, this test is only designed to test the null hypothesis (no difference) against a shift alternative. For practical application, arithmetic simplifications can be used. This may hide the relation between the usual formula for the t-test and for the Wilcoxon test.

To apply the Wilcoxon test, on the one hand the test statistic has to be calculated. On the other hand, to determine the critical values, the distribution function has to be evaluated. If all observations are distinct, this function depends only on n_1 and n_2, and fairly simple algorithms are available. These are provided in the R base and used by `wilcox.test()`. However, if there are ties in the data, that is, there are values occurring more than once, the distribution depends on the special pattern of these ties and the calculation is laborious. `wilcox.test()` returns to approximations in this case. For an exact evaluation (in contrast to approximative), the necessary algorithms are available as well. To use them, you need `library(coin)`. The exact variant of the Wilcoxon tests, for example, is implemented as `wilcox_test()`.

It is a classical branch in statistics to characterise distribution-independent methods based on ranks and to compare them with independent methods based on Monte Carlo calculations and their variations. You find literature for this area under the keywords "rank tests" and "distribution free methods". Additional R functions are provided in `libary(coin)` as well as in some specialised libraries.

Of course information loss is a concern. If we restrict ourselves to the data and make no or weak assumptions on the distributions, we have less information than in a model with explicit distribution assumptions. If we reduce the data to ranks, we may give away additional information. This loss in information can be measured, for instance, in terms of asymptotic relative efficiency, that is, the (asymptotic) sample size proportion that is needed to get a comparative power using the competing test. Under the assumption of normality, the asymptotic relative efficiency of the Wilcoxon test is 94%. So if the assumption of a normal distribution holds, the (optimal) t-test only needs 94% of the sample size needed by the the Wilcoxon test. 6% of the sample size is the costs for the reduction to ranks. But if the normal distribution does not apply, the t-test possibly can break down, while the Wilcoxon test stays valid for the shift alternative.

---

help(wilcox.test)

`wilcox.test` *Wilcoxon Rank Sum and Signed Rank Tests*

Description

Performs one and two sample Wilcoxon tests on vectors of data; the latter is also known as 'Mann-Whitney' test.

Usage

```
wilcox.test(x, ...)

## Default S3 method:
wilcox.test(x, y = NULL,
            alternative = c("two.sided", "less", "greater"),
            mu = 0, paired = FALSE, exact = NULL, correct = TRUE,
            conf.int = FALSE, conf.level = 0.95, ...)

## S3 method for class 'formula':
wilcox.test(formula, data, subset, na.action, ...)
```

Arguments

`x`	numeric vector of data values. Non-finite (e.g. infinite or missing) values will be omitted.
`y`	an optional numeric vector of data values.
`alternative`	a character string specifying the alternative hypothesis, must be one of `"two.sided"` (default), `"greater"` or `"less"`. You can specify just the initial letter.
`mu`	a number specifying an optional parameter used to form the null hypothesis. See 'Details'.
`paired`	a logical indicating whether you want a paired test.

`exact`	a logical indicating whether an exact p-value should be computed.
`correct`	a logical indicating whether to apply continuity correction in the normal approximation for the p-value.
`conf.int`	a logical indicating whether a confidence interval should be computed.
`conf.level`	confidence level of the interval.
`formula`	a formula of the form `lhs ~ rhs` where `lhs` is a numeric variable giving the data values and `rhs` a factor with two levels giving the corresponding groups.
`data`	an optional matrix or data frame (or similar: see `model.frame`) containing the variables in the formula `formula`. By default the variables are taken from `environment(formula)`.
`subset`	an optional vector specifying a subset of observations to be used.
`na.action`	a function which indicates what should happen when the data contain `NA`s. Defaults to `getOption("na.action")`.
`...`	further arguments to be passed to or from methods.

Details

The formula interface is only applicable for the 2-sample tests.

If only `x` is given, or if both `x` and `y` are given and `paired` is `TRUE`, a Wilcoxon signed rank test of the null that the distribution of `x` (in the one sample case) or of `x - y` (in the paired two sample case) is symmetric about `mu` is performed.

Otherwise, if both `x` and `y` are given and `paired` is `FALSE`, a Wilcoxon rank sum test (equivalent to the Mann-Whitney test: see the Note) is carried out. In this case, the null hypothesis is that the distributions of `x` and `y` differ by a location shift of `mu` and the alternative is that they differ by some other location shift (and the one-sided alternative `"greater"` is that `x` is shifted to the right of `y`).

By default (if `exact` is not specified), an exact p-value is computed if the samples contain less than 50 finite values and there are no ties. Otherwise, a normal approximation is used.

Optionally (if argument `conf.int` is true), a nonparametric confidence interval and an estimator for the pseudomedian (one-sample case) or for the difference of the location parameters `x-y` is computed. (The pseudomedian of a distribution F is the median of the distribution of $(u + v)/2$, where u and v are independent, each with distribution F. If F is symmetric, then the pseudomedian and median coincide. See Hollander & Wolfe (1973), page 34.) If exact p-values are available, an exact confidence interval is obtained by the algorithm described in Bauer (1972), and the Hodges-Lehmann estimator is employed. Otherwise, the returned confidence interval and point estimate are based on normal approximations.

With small samples it may not be possible to achieve very high confidence interval coverages. If this happens a warning will be given and an interval with lower coverage will be substituted.

Value

A list with class `"htest"` containing the following components:

`statistic`	the value of the test statistic with a name describing it.
`parameter`	the parameter(s) for the exact distribution of the test statistic.

`p.value`	the p-value for the test.
`null.value`	the location parameter `mu`.
`alternative`	a character string describing the alternative hypothesis.
`method`	the type of test applied.
`data.name`	a character string giving the names of the data.
`conf.int`	a confidence interval for the location parameter. (Only present if argument `conf.int = TRUE`.)
`estimate`	an estimate of the location parameter. (Only present if argument `conf.int = TRUE`.)

Warning

This function can use large amounts of memory and stack (and even crash R if the stack limit is exceeded) if `exact = TRUE` and one sample is large (several thousands or more).

Note

The literature is not unanimous about the definitions of the Wilcoxon rank sum and Mann-Whitney tests. The two most common definitions correspond to the sum of the ranks of the first sample with the minimum value subtracted or not: R subtracts and S-PLUS does not, giving a value which is larger by $m(m+1)/2$ for a first sample of size m. (It seems Wilcoxon's original paper used the unadjusted sum of the ranks but subsequent tables subtracted the minimum.)

R's value can also be computed as the number of all pairs `(x[i], y[j])` for which `y[j]` is not greater than `x[i]`, the most common definition of the Mann-Whitney test.

References

David F. Bauer (1972), Constructing confidence sets using rank statistics. *Journal of the American Statistical Association* **67, 687–690.**

Myles Hollander & Douglas A. Wolfe (1973), *Nonparametric Statistical Methods.* New York: John Wiley & Sons. Pages 27–33 (one-sample), 68–75 (two-sample).
Or second edition (1999).

See Also

`psignrank`, `pwilcox`.

`wilcox.exact` in **exactRankTests** covers much of the same ground, but also produces exact p-values in the presence of ties.

`wilcox_test` in package **coin** for exact and approximate *conditional* p-values for the Wilcoxon tests.

`kruskal.test` for testing homogeneity in location parameters in the case of two or more samples; `t.test` for an alternative under normality assumptions [or large samples]

Examples

```
require(graphics)
## One-sample test.
## Hollander & Wolfe (1973), 29f.
## Hamilton depression scale factor measurements in 9 patients with
##  mixed anxiety and depression, taken at the first (x) and second
##  (y) visit after initiation of a therapy (administration of a
##  tranquilizer).
x <- c(1.83,  0.50,  1.62,  2.48, 1.68, 1.88, 1.55, 3.06, 1.30)
y <- c(0.878, 0.647, 0.598, 2.05, 1.06, 1.29, 1.06, 3.14, 1.29)
wilcox.test(x, y, paired = TRUE, alternative = "greater")
wilcox.test(y - x, alternative = "less")    # The same.
wilcox.test(y - x, alternative = "less",
            exact = FALSE, correct = FALSE) # H&W large sample
                                            # approximation

## Two-sample test.
## Hollander & Wolfe (1973), 69f.
## Permeability constants of the human chorioamnion (a placental
##  membrane) at term (x) and between 12 to 26 weeks gestational
##  age (y).  The alternative of interest is greater permeability
##  of the human chorioamnion for the term pregnancy.
x <- c(0.80, 0.83, 1.89, 1.04, 1.45, 1.38, 1.91, 1.64, 0.73, 1.46)
y <- c(1.15, 0.88, 0.90, 0.74, 1.21)
wilcox.test(x, y, alternative = "g")        # greater
wilcox.test(x, y, alternative = "greater",
            exact = FALSE, correct = FALSE) # H&W large sample
                                            # approximation

wilcox.test(rnorm(10), rnorm(10, 2), conf.int = TRUE)

## Formula interface.
boxplot(Ozone ~ Month, data = airquality)
wilcox.test(Ozone ~ Month, data = airquality,
            subset = Month %in% c(5, 8))
```

Exercise 3.16	
	Use the Wilcoxon test to compare the results of the right/left `click` experiment.

Exercise 3.17	Click Project
***	In the the right/left `click` experiment several effects contribute to the response time. Some problems: • The response time comprises reaction time, time for the large scale movement of the mouse, time for fine adjustment, etc. • For the right/left movement in general a swivel of the hand is sufficient. For forward/backward movement in general a movement of the arm is necessary. It is not to be expected that both movements have a comparable statistical behaviour. • Subsequent records may be affected by a training effect, or by a tiring effect. Can you modify the experiment or the evaluation so that differences in the reaction time components can be investigated? Can you modify the experiment or the evaluation so that differences in the precision of the position of the click can be investigated?
***	Inspect and document for yourself the right/left differences in reaction time and precision. Summarise your results as a report.

Exercise 3.18	Power Comparison
	Use the shift/scale families of $N(0,1)$ and $t(3)$ and design a setting to compare the performance of the Wilcoxon test with that of the t-test for each of these families. Perform the comparison in a simulation with sample sizes $n_1 = n_2 = 10, 20, 50, 100$ and summarise your results.
	Do an analogous comparison using simulation data from the lognormal distribution.

3.4 A Road Map

If you go hiking, a road map may be on the middle of the way. Most information is also on the map that you may carry with you, but the essential additional information is a point saying "You Are Here". We are in a similar situation. At this point, we can outline the road we have taken, and the perspectives ahead.

We started with basic data analysis in Chapter 1. This chapter discussed independent, identically distributed data; however, the discussion was still in a void space, missing a clear target.

The linear models in Chapter 2 gave a well-defined framework, and put Chapter 1 in a context. We need basic data analysis, but if we follow the approach of linear models we need diagnostics based on residuals. For linear models, the stochastic part is in the error term. But the stochastic errors are not directly observable. We have to resort to an indirect inference, based on the residuals. We have two ways we can go. We can derive diagnostic indices based on the residuals, which are at least approximatively independent, identically distributed data and use basic analysis as in Chapter 1. Or we can modify the basic analysis to cope with the dependent inhomogeneous structure derivable from the data. Both approaches are feasible. And both approaches have to cope with the problem that the residuals are affected both by the data and by our estimation step. So we have to avoid running in circles, such as hidden outliers or masking effects.

In this chapter, we have put linear models in their place. For the simple two-sample situation, linear models boil down to comparison of two treatments. What they provide is analysis for a shift. Formally, linear models consider a shift/scale family, but at least for simple linear models the scale is just a nuisance parameter that is removed as soon as possible. If you have some good reasons to believe in the distribution assumptions, linear models are a first choice. More often, you may be working in a field where experimental set-ups and appropriate scale transformations are well known enough to at least believe in the distribution assumptions with good confidence. (Be aware of the pitfalls. The assumptions or transformations to enforce the assumptions may be well established for a "normal" population. But are they established for a population to compare with, for example, under treatment?) If the assumptions can be made, fine, but use diagnostics as a second check. If they are in doubt, you can use non-parametric methods at a negligible cost, keeping competitive performance when the distribution assumptions are satisfied, but guaranteeing quality even if the distribution assumptions fail. The standard methods, however, are only tuned for shift alternatives.

The main message of this chapter is that the world is richer than what shift alternatives suggest. Be prepared for the unexpected.

The other open horizon is to go to higher dimensions. So far, the data have been one dimensional. Well, not quite. The data in Chapter 2 were essentially $p+1$-dimensional. It was the specific view of the regression model that factorized the data into a regressor part X of dimension p, which is considered non-stochastic, and a response part Y, which was modelled as stochastic, and of course it is the question to the application whether this split is adequate, or whether the data should be analysed as a joined higher-dimensional data set. Dimensionality is a topic still to come. This topic will be touched on in Chapter 4.

With this said, we can return to our familiar procedures. We will have a closer look at their power, and see how this information is related to practical planning of experiments.

3.5 Power and Confidence

3.5.1 Theoretical Power and Confidence

We can use the t-test to illustrate how tests are constructed in general. The test uses a statistic, here the t-test statistic for comparison of two groups (3.1). We know the distribution of this statistic. In the case of the t-test, given independent errors with normal distribution and equal variance the test statistic has a $t(n_1+n_2-2)$ distribution. For a given level α we can use the distribution function to read off limits that are exceeded resp. missed only with a probability of α. If we use both limits we get a two-sided region with error probability 2α.

***Example* 3.4: Test Construction**

Input

```
n1<- 6; n2 <- 6
df <-  n1 + n2 -2
alpha <- 0.05
curve(pt(x,df=df),from=-5, to=5, ylab= expression(F[n]),
    main="t-Test: Critical Value")
abline(h=1-alpha, col="red")       # cut at upper quantile
abline(v=qt(1-alpha, df=df), lty=3, col="red") # get critical value
legend("topleft", legend=c("level","critical value"),
    lty=c(1,3),col="red",
    bty="n", inset=c(0,0.2))
```

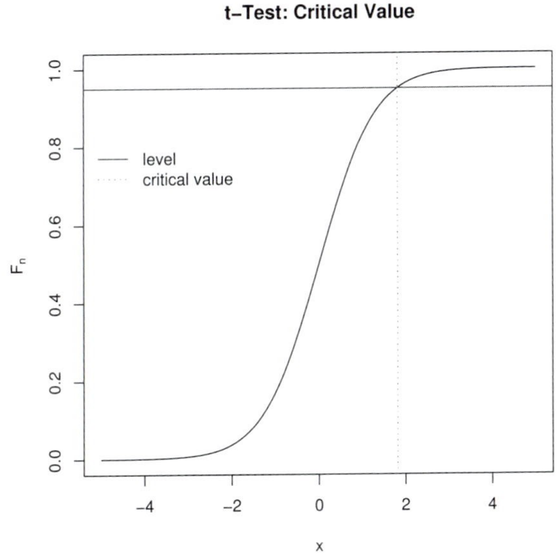

If, for example, we want to test the hypothesis $\mu_1 = \mu_2$ against the alternative $\mu_1 > \mu_2$, we use the region above the upper of these limits as a rejection region. We know that if the hypothesis applies, a random test statistic will fall into this region with probability of at most α.

For the t-test we know more. Under the model assumptions of independent errors with normal distribution and equal variance, the t-test statistic always has a t distribution. On the hypothesis it has a t distribution with non-centrality parameter 0, that is, it has a central t distribution. On the alternative where $\mu_1 \neq \mu_2$, but still for normal distribution with equal variance, we have a t distribution with non centrality parameter $(\mu_1 - \mu_2)\sigma^{-1}\sqrt{n_1 n_2/(n_1 + n_2)}$. Under model assumptions this gives the power of the test for each alternative, that is, the probability to reject the hypothesis if a specific alternative applies.

Example **3.5: Power (Fixed Non-Centrality)**

Input

```
n1<- 6; n2 <- 6
df <-  n1 + n2 -2
alpha <- 0.05
curve(pt(x,df=df),from=-5, to=5, ylab= expression(F[n]),
    main="Central and Non-Central t-Distribution")
abline(h=1-alpha, col="red")       # cut at upper quantile
abline(v=qt(1-alpha, df=df), lty=3, col="red") # get critical value
n1 <- 5
n2 <- 5
n <- n1+n2
theta <- 2
ncp <- theta * sqrt(n1 * n2/(n1+n2))
mtext(paste("non-centrality",round(ncp,2)))
curve(pt(x,df=df, ncp=ncp), lty=2, add=TRUE, col="blue")
legend("topleft", legend=c("central t","non-central t"),
    lty=c(1,2), col=c("black","blue"),
    bty="n", inset=c(0,0.2))
```

Central and Non-Central t-Distribution

non-centrality 3.16

central t
non-central t

F_n

x

We can display the power.[2] of the test by plotting the rejection probability as a function of $\frac{\mu_1-\mu_2}{\sigma}$.

[2] Conventionally, the term "power function" is only used for the rejection probability on the alternative, that is, for example, for $(\mu_1 - \mu_2) > 0$.

***Example* 3.6: Power Function**

Input

```
tpower <- function(n1, n2, alpha,...){
        df <-  n1 + n2 -2
        tlim <- qt(1-alpha,df=df)
        prob <- function(theta){
            pt(tlim, df = df,
                ncp = theta * sqrt(n1 * n2/(n1+n2)),
                lower.tail=FALSE)}
        curve(prob, 0, 5, xlab=expression(theta==mu[1]-mu[2]), ...)
        abline(h=alpha, col="red")
}

tpower(5, 5, 0.05, main="Power Function for Selected Sample Sizes")
tpower(10, 10, 0.05, add =TRUE, lty = 3)
tpower(100,100, 0.05, add =TRUE, lty = 4)
tpower(1000, 1000, 0.05, add =TRUE, lty = 5)
legend("bottomright",
    lty=c(1,3,4,5),
    legend=c("n1 = n2 =5", "n1 = n2 =10", "n1 = n2 =100","n1 = n2 =1000"),
    inset=0.1, bty="n")
```

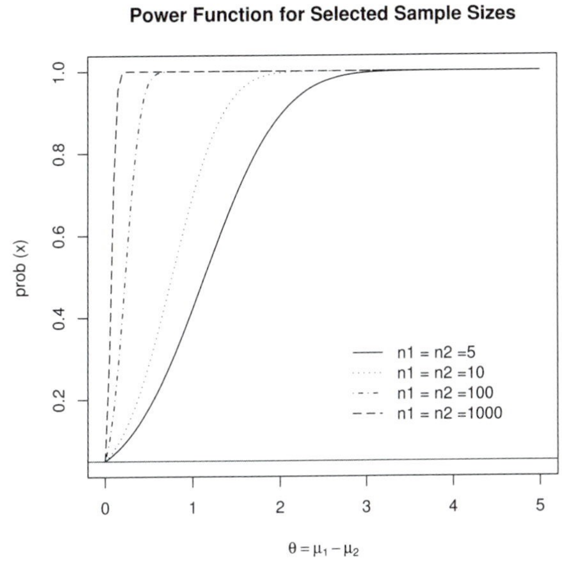

This relation can also be used to calculate the sample size needed to reject the hypothesis with a probability α if it applies (false rejection), but to reject the hypothesis if some specified alternative applies (power or correct rejection). Function `power.t.test()` is already prepared to do this calculation.

Example 3.7: Sample Size Determination

Input

```
power.t.test(delta=2,
    power=0.8,
    sig.level=0.01,
    type="two.sample",
    alternative="one.sided")
```

Output

```
    Two-sample t test power calculation

              n = 6.553292
          delta = 2
             sd = 1
      sig.level = 0.01
          power = 0.8
    alternative = one.sided

NOTE: n is number in *each* group
```

3.5.2 Simulated Power and Confidence

If the theoretical properties of a test are known, this is the best way to analyse its properties. In an environment such as R we have the possibility of assessing the power even if theoretical results are not available or not accessible. For chosen alternatives we can generate random samples, perform tests and find the relative proportion of samples that lead to a rejection. If we generate *nsimul* independent random samples with identical distribution, the number of rejections is a random variable with binomial distribution and

$$\widehat{p} = \frac{\#rejections}{nsimul}$$

is an estimator for the rejection probability.

As an example we see how the t-test performs if the data have a log-normal distribution. We compare two groups, each with sample size $n_1 = n_2 = 10$. We generate data under the hypothesis first:

***Example* 3.8: Violation of Assumptions (log-normal)**

Input

```
nsimul <- 500
n1<- 10; n2 <- 10
alpha <- 0.01 #nominal level
x <- 0
for (i in 1:nsimul) {
        if (t.test(exp(rnorm(n1)),exp(rnorm(n2)),
        alternative="less",
        var.equal = TRUE)$p.value < alpha){
            x <- x+1}
         }
p <- x/nsimul
cat("estimated level p", p)
```

Output

```
estimated level p 0.006
```

However, there is an open issue here. We have used just 500 simulations. So for a nominal level of $\alpha = 0.01$ we would expect just 5 rejections. Our simulation result can be heavily influenced by random effects. The function *prop.test()* calculates not only the estimator for the binomial fraction, but also a confidence set.

***Example* 3.9: Binomial Confidence Set**

Input

```
prop.test(n=nsimul, x=x)
```

Output

```
        1-sample proportions test with continuity correction

data:  x out of nsimul, null probability 0.5
X-squared = 486.098, df = 1, p-value < 2.2e-16
alternative hypothesis: true p is not equal to 0.5
95 percent confidence interval:
 0.001550854 0.018951807
sample estimates:
    p
0.006
```

***Example* 3.10:**

Input

```
nsimul <- 500
n1<- 10; n2 <-10
alpha <- 0.01
x<-0
for (i in 1:nsimul) {
        if (t.test(exp(rnorm(n1)),exp(rnorm(n2, mean = 1)),
            alternative="less",
            var.equal = TRUE)$p.value < alpha){
                x <- x+1}
        }
p <- x/nsimul
cat("estim p", p)
```

Output

```
estim p 0.188
```

Input

```
prop.test(n = nsimul, x = x)
```

Output

```
        1-sample proportions test with continuity correction

data:  x out of nsimul, null probability 0.5
X-squared = 193.442, df = 1, p-value < 2.2e-16
alternative hypothesis: true p is not equal to 0.5
95 percent confidence interval:
 0.1552568 0.2256388
sample estimates:
    p
0.188
```

In `library(binom)` [8] several tools are provided for a differentiated analysis of the binomial distribution.

The confidence intervals in this example show that a simulation sample size of $nsimul = 500$ gives only rough results. For simulations, we want better control over the simulation precision. We can choose the simulation sample size to improve the precision as we like. A sample size calculation can be done with `power.prop.test()`. In experiments, an exact sample size calculation may be necessary. For simulations, often an estimation of the sample size is sufficient.

With $\widehat{p} := Z/n$ as estimator for a probability p we have $E(\widehat{p}) = p$ and $Var(\widehat{p}) = p(1-p)/n$. If p is actually the value of the parameter, an error has to be judged relative to this target parameter. At $p = 50\%$ an error of $\pm 1\%$ has to be judged differently than at $p = 99\%$. The relative error, the ***coefficient of variation***, is

$$\frac{\sqrt{Var(\widehat{p})}}{E(\widehat{p})} = \sqrt{\frac{1-p}{np}}.$$

To get a coefficient of variation of at most η, we need a sample size

$$n \geq \frac{1-p}{p\eta^2}.$$

If n and p are in an order of magnitude where a normal approximation applies, we have an approximate confidence interval with limits

$$\widehat{p} \pm \phi_{1-\alpha/2}\sqrt{\frac{\widehat{p}(1-\widehat{p})}{n}} \text{ for a confidence level } 1-\alpha.$$

If the length of the confidence interval should not exceed ηp, we need a sample size

$$n \geq \frac{\phi^2_{1-\alpha/2}(1-p)}{p\eta^2}.$$

As usual, the choice of α is up to us. For example, for α =1% with $\phi_{1-\alpha/2} = 2.575829$ we get the values in Table 3.29. If we work with higher quantiles, we will try to bound the error relative to $1-p$. Examples are in Table 3.29.

p	$1-p$	$n(\alpha = 10\%)$		$n(\alpha = 1\%)$	
		$\eta = 0.1$	$\eta = 0.01$	$\eta = 0.1$	$\eta = 0.01$
0.500	0.500	271	27 055	663	66 349
0.250	0.750	812	81 166	1990	199 047
0.100	0.900	2435	243 499	5971	597 141
0.010	0.990	26 785	2 678 488	65 685	6 568 547
0.001	0.999	270 283	27 028 379	662 826	662 822 617

Table 3.29 *Required sample size for two-sided confidence intervals with relative length* $\leq \eta$

It is worth memorising the rough numbers: To estimate a probability in the range of magnitude of $50\% \pm 5\%$ with 90% confidence, about 300 simulations are necessary. To estimate a value of about 99% up to $\pm$ 0.1%, we need 30000 simulations.

Another issue worth noting here is that the required sample size depends drastically on the probability level at which we are working. If we can design our experiment or simulation so that we are not working in the tails, but near to the centre of the distribution (for example, by using some appropriate conditional distribution), we can reduce the required sample size from some ten or hundred thousands to just some hundreds. Techniques to achieve this are discussed in the literature under the topic ***importance sampling***, or ***cheating*** as it is called in the earlier literature.

3.5.3 Quantile Estimation

The other side of the problem above is to estimate a quantile from a random sample. We already know that, for a random variable X with continuous distribution function F, the variable

$F(X)$ has a uniform distribution on $[0,1]$. For the quantile estimation we need the distribution function, evaluated at the order statistics.

Theorem 3.2 *For independent observations $X_i, i = 1, \ldots, n$ from a continuous distribution function F, let $X_{(k:n)}$ denote the k-th order statistic. Then*

$$F(X_{(k:n)}) \sim P_{beta}(\,\cdot\,; k, n-k+1).$$

Proof. [38] or [57], 8.7.2 □

We repeat:

Remark 3.3 *In general the beta distribution is skewed. The expected value of the beta$(k, n-k+1)$ distribution is $k/(n+1)$. For an unbiased estimation of the quantile x_p use $X_{(k:n)}$ with $k/(n+1) = p$. The "plug in" approximation $k/n = p$ gives a biased estimation.*

The theorem can be applied directly to get an upper or lower estimate for quantiles. As a special case we can try to use the minimum of the observed values $X_{(1:n)}$ as lower estimate for the p-quantile. The confidence level is

$$P(X_{(1)} \leq F_p) = P(F(X_{(k)}) \leq p) = I_p(1, n),$$

where I is the incomplete beta integral. For the special parameters $(1, n)$ the beta density simplifies to $n(1-p)^{n-1}$ and we get the incomplete beta integral $I_p(1,n) = 1-(1-p)^n$. Hence

$$P(X_{(1)} \leq F_p) = 1-(1-p)^n$$

and we can a guarantee a confidence level $1-\alpha$, if

$$n \geq \frac{\ln \alpha}{\ln(1-p)}.$$

Similarly, the observed maximum value can be used as an upper estimate for the p-quantile. By symmetry we get a confidence level of $1-\alpha$, if

$$n \geq \frac{\ln \alpha}{\ln p}.$$

Examples are given in Table 3.30.

p		n			
$X_{(1)} \leq F_p$	$X_{(n)} \geq F_p$	$\alpha = 10\%$	$\alpha = 5\%$	$\alpha = 1\%$	$\alpha = 0.5\%$
0.500	0.500	4	5	7	8
0.250	0.750	9	11	17	19
0.100	0.900	22	29	44	51
0.010	0.990	230	299	459	528
0.001	0.999	2302	2995	4603	5296

Table 3.30 *Required sample size for estimation of a quantile with confidence level $\geq 1-\alpha$*

Again it is worth memorising the rough numbers: to get a one-sided estimate for a 1% (99%)-quantile of a continuous distribution function with a confidence of 99%, about 500 simulations are needed.

We can combine one-sided limits to two-sided intervals. The corresponding result to calculate the probability of intervals is in Corollary 3.4:

Corollary 3.4 *For the k_1th and $k_1 + k_2$th order statistic, the interval $(X_{(k_1:n)}, X_{(k_1+k_2:n)})$ is a confidence interval for the p-quantile with coverage probability*

$$I_p(k_1, n - k_1 + 1) - I_p(k_1 + k_2, n - k_1 - k_2 + 1).$$

Proof. [57] 11.2.1 □

The simulation sample sizes for quantile estimation are considerably smaller than those needed for the estimation of comparable probabilities. In retrospect, this is not surprising: the question whether an observation exceeds some specific quantile is simpler then the task to estimate the p-value. The sample size can often be reduced even more drastically, if the question at hand is reduced to a test problem.

Without additional distribution assumptions this gives a first possibility to choose the size of a simulation. In special situations, clever ideas can allow for a drastic reduction of the simulation sample size. But for a start, the calculations given above provide a basis for determining the simulation sample size.

3.6 Qualitative Features of Distributions

At this point, we again ask about the purpose of data analysis. Chapter 1 was quite open to questioning this. We met some traditional aspects, such as getting information about quantiles or moments. From a more general point of view, we encountered questions about location and scale. Moments or quantiles found their place as intermediates to derive this information of interest.

Chapter 2 led to a more focussed view. On the one hand, specific parameters such as the regression coefficients and the scale parameter of the error distribution appeared as central targets. On the other hand, indicators for deviations from the linear model or trouble makers such as effective outliers were of interest. This is a typical application example that leads to an analysis of the data which must be linked to the intended statistical analysis of the data, and hence to the intended evaluation method.

Focus has changed with available technology. In the first half of the last century, linear models were the prominent method that could be covered computationally, and the steps were simple. First, try to get hold of the first two moments (mean and variance of your data). Second, try to find a transformation such that in the framework of a linear model the next two moments of the estimated error are approximitavely those of a normal distribution. So the third moment, the skewness, should be zero after transformation. The fourth was compared to that of a normal distribution, yielding a kurtosis excess, which should be transformed to zero. Then a linear model would be applied.

Technology has advanced, more computing power is available, and occasionally we have more advanced algorithms. So this "method of the moments" has implicitly migrated into estimators and lost importance.

We have a remainder of this point of view in the box-and-whisker plot. But it goes beyond the traditional view. It uses the median as a location estimator, but has the upper and lower quartile yielding possibly asymmetric scale estimators to the upper and lower range. The information about out and far-out points is additional information that was not considered in earlier approaches.[3]

On the other hand, we have what can be considered density estimators. The histograms are early representatives of this family. Kernel density estimators are followers. Both seem to give an intuitive impression of the distribution. But this view changes with different settings. There is a bandwidth, or smoothing parameter, which critically affects the impression. Neither yields easily for a multi-scale representation.

The problem does not seem to lie with the method, but with the application. Which features are relevant for the application?

Theory can deliver solutions for standard problems, such as characterisation of location and scale. It may be helpful to have a comparison basis. Now assume that location and scale are analysed and have been adjusted. What is the next step for comparison?

If we have a shift/scale family, all is solved now. If we have a stochastic ordering, comparison of distribution functions may help. But a typical example may just shift a part of the distribution.

A particular case may arise where you have a distinct group of responders. So in the two-sample case, a possibly unimodal distribution in the control population may yield a bimodal distribution under treatment. The effect of interest in this case is to detect bimodality. A simple solution is to use the shorth functional: for any data point x and any chosen level α find the length of the smallest interval containing x and covering at least a proportion α of the data. You can draw this length as a function of x for various levels α since the curves are not overlapping. So you do not run into the bandwidth problem of histograms or kernel density estimators. We will use this idea in the next chapter (Section 4.7.3 (page 173)).

3.7 Statistical Summary

As a leading example in this chapter we use the comparison of treatments. In simple cases, the samples under the treatments to be compared differ only by a shift in mean. In this case the problems can be reduced to the approaches presented in Chapter 2. In this reduced case, the distributions coincide when centred at the mean. For the general case that we touched on here, the simplification does not work. An important example is the analysis of therapy studies. If a treatment has a homogenous effect, we can analyse it using the methods in Chapter 2. But often under treatment we see a split into "responders" and "non-responders", or we get a qualitatively different distribution under treatment than in the control group. This goes beyond the models sketched in Chapter 2 and requires more general approaches mentioned in this chapter.

We have restricted our discussion to the comparison of two samples. Practical work often leads to other problems. For example, a typical question is to compare a new treatment with a known reference treatment, where only sample observations are available for the new treatment, but extensive prior information is available about the reference treatment. Or a reference treatment is to be compared with a series of alternative treatments. These problems go beyond the scope of our introduction. Here we can only refer to advanced literature, for example, [28].

[3] The original proposal for the definition of out and far-out points by J. Tukey just reflects typical sample sizes at the time when this proposal was made. For large or small sample sizes, it needs some improvement.

3.8 Literature and Additional References

[52] William N. Venables and Brian D. Ripley, B (2002): *Modern Applied Statistics with S.* Springer, Heidelberg.

[51] William N. Venables, W.N.; and Brian D. Ripley (2000): *S Programming.* Springer, Heidelberg.

[28] Rupert G. Miller (1981): *Simultaneous Statistical Inference.* Springer, Heidelberg.

CHAPTER 4

Dimensions 1, 2, 3, ..., ∞

In the previous chapters, we used numerical and graphical tools in parallel. The general idea is that there is a strict correspondence. A graphical tool is only worth as much as the hard statistics that is supporting it. If we go to higher dimensions, we have to review this correspondence.

Some of the problems with higher dimensions come from perception. It is doubtful that humans are made for handling numbers. Even if we are numerate, most of us barely can go beyond one dimension as far as numbers are concerned. We can display numbers in a table, generally with a matrix layout. This gives us two dimensions for the position, and one for the value. But if you make any experiments, you will notice that few individuals are able to exploit this three-dimensional structure. Multi-way tables are even more of a challenge to read. Numbers are for one-dimensional people.

Graphical representations are more accessible in a higher-dimensional setting than numerical representations, at least if the graphical structures correspond to those familiar from natural phenomena. Most people are able to see evolving three-dimensional structures. So four dimensions (three for the space, one for time) seem to be tractable, and most people are capable of tracing one or two qualitative features (such as colour or texture) along with the basic geometry, giving a perceptual space of four to six dimensions to start with.

While it may be interesting to investigate quantitative representations in higher dimensions, we will focus on graphical aspects in this chapter.

Graphics is related to geometry. But in a statistical context, they are not the same. Each variable may have its own scale, and these scales may have dependencies. Finding an appropriate geometry is a topic in itself. A first step in this direction is to find the proper scale of variables. In the following, we will often assume that a proper scaling has been done. This leaves a large open space that is not covered here.

To give an impression of the dragons that are lurking out there, we mention just two problems that are not covered here. Assume you have a data set with n observations $X_i, i = 1, \ldots, n$ each recorded formally with a p-dimensional variate. Assume that you have a measure of distance $d(X_i, X_j)$ for any two data points. Find a geometric representation of the data in a q-dimensional variate, for example a planar representation in two-space. You will find approaches to this problem under the keyword ***multidimensional scaling***. A whole branch of statistics is related to this kind of problems. The monograph [3] is a standard textbook in this field. We will only touch on it and assume that a proper scaling has been done.

The other problem that already may occur in one dimension, but is more typical for higher dimensions, is to find groups in data. This occurs in two variants. One is to start with defined groups and to derive the rules defining these groups. In a statistical context, this is usually called a classification problem. In computer science, this is often called "supervised learning". The other variant is to find groups, based on the information contained in the data. In a statistical context, this is called clustering. Computer scientists use the term "unsupervised learning". A rich monograph on these topics is [14].

When going to higher dimensions, we have to cover two aspects. On the one hand, we have to study the new features that appear in higher dimensions that go beyond what is familiar in lower dimensions. On the other hand, we have to study the possibilities to use lower-dimensional tools for the investigation of higher dimensions. We will start our discussion with a look at some of the standard tools.

4.1 R Complements

In this chapter we begin with complements on R in order to concentrate on statistical questions for the remainder without disrupting the discussion with programming details. We take a look at the graphical possibilities that are at our disposal. The basic graphics model of R is oriented to possibilities historically provided by a plotter as an output device. The graphics follow the possibilities available when drawing with a pen. Besides the one- and two-dimensional possibilities which we have seen so far, there are possibilities to display a real valued function that is defined over a grid. Basically, three R functions are available for this.

3d Basic Graphics	
`image()`	gives the values of a variable `z` in grey levels or colour coding.
`contour()`	gives the contours of a variable `z`.
`persp()`	gives a perspective plot of a variable `z`.

The basic graphics system in R is easy to use, but limited in possibilities. A newer graphics system, the grid and lattice graphics [41], conceptually works with objects and a viewport model. The graphic objects can be combined and post-processed. The display takes part in a separate step. Simple 2d graphs can be post-processed. For a 3d display, distance, the point of view and the focal length can be chosen as we would do when using a camera. The object-oriented graphics system consists of a library `grid` with the elementary operations required, and a higher level library `lattice` that gives new implementation of the displays known from the basic graphics and adds additional displays.

`image()` and `contour()` can also be used to give an overlay on other plots.

***Example* 4.1: 3d Surface Displays Using Base Graphics**

Input

```
x <- 10*(1:nrow(volcano))
y <- 10*(1:ncol(volcano))
image(x, y, volcano, col = terrain.colors(100),
    axes = FALSE, xlab = "Meters North", ylab = "Meters West")
axis(1, at = seq(100, 800, by = 100))
axis(2, at = seq(100, 600, by = 100))
box()
title(main = "Maunga Whau Volcano", font.main = 4)

contour(x, y, volcano, levels = seq(90, 200, by = 5),
         col = "peru", main = "Maunga Whau Volcano", font.main = 4,
         xlab = "Meters North", ylab = "Meters West")
z <- 2 * volcano         # Exaggerate the relief
x <- 10 * (1:nrow(z))    # 10 meter spacing (S to N)
y <- 10 * (1:ncol(z))    # 10 meter spacing (E to W)

## Don't draw the grid lines :  border = NA
persp(x, y, z, theta = 135, phi = 30, col = "green3", scale = FALSE,
      ltheta = -120, shade = 0.75, border = NA, box = FALSE)
```

`image()` `contour()` `persp()`

See Colour Figure 3.

3d Lattice Graphics	
`cloud()`	generic lattice function to draw 3d scatterplots.
`wireframe()`	generic lattice function to draw 3d surfaces.

In the basic graphics system, the functions usually provide a graphical output, and the internal information must be accessed explicitly. In the lattice system, the functions usually return lattice objects. Graphical output must be requested explicitly. For the output of lattice objects the function *print()* is used.

***Example* 4.2: 3d Surface Display Using Lattice Graphics**

Input

```
library(lattice)
print(wireframe(volcano, shade = TRUE,
        aspect = c(61/87, 0.4),    ## volcano  ## 87 x 61 matrix
        par.settings = list(axis.line = list(col = "transparent")),
        light.source = c(10,0,10)))
```

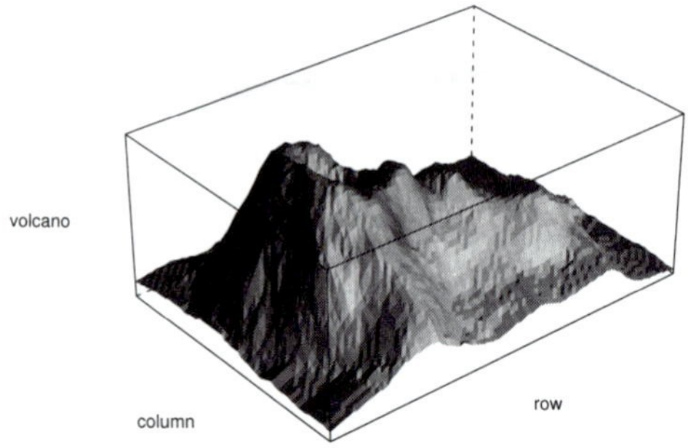

volcano
column
row

See Colour Figure 4.

If you know that you are displaying 3d scenes, you might consider `library(rgl)` [1] as an alternative. If implemented on your system, `rgl` provides real-time 3d rendering with interactive facilities. This code snippet will allow you to turn the vulcano upside down:

```
library("rgl")
example(surface3d)
```

The basic graphics system and lattice graphics are separate graphics systems. Unfortunately, they use different notations for comparable functions, and comparable displays have different representations. A small translation aid is given in Table 4.5. Some convenience functions to combine both graphics systems are provided in library *gridBase*. An extensive introduction to both graphics systems is [29].

In a wide range of scientific visualisations, OpenGL is used as a common standard. OpenGL

Basic Graphics		Lattice
barplot()	bar chart	*barchart()*
boxplot()	box-and-whisker plot	*bwplot()*
	three-dimensional scatterplot	*cloud()*
contour	contour plot	*contourplot()*
coplot	conditional scatterplots	*xyplot()*
plot(density())	density estimator	*densityplot()*
dotchart()	dot plot	*dotplot()*
hist()	histogram	*histogram()*
image()	colour map plots	*splom()*
	parallel coordinate plots	*parallel()*
pairs()	scatterplot matrices	*wireframe()*
persp()	three-dimensional surface	*wireframe()*
plot()	scatterplot	*xyplot()*
qqnorm()	theoretical QQ plot	*qqmath()*
qqplot()	empirical QQ plot	*qq()*
stripchart()	one-dimensional scatterplot	*stripplot()*

Table 4.5 *Basic graphics and lattice graphics*

functions are accessible in R using the library *rgl*. There are, however, certain differences between common requirements for graphics, and the specific requirements of statistical graphics. As far as the representation of functions is concerned, statistical graphics is comparable with the requirements usual in analysis. The small difference is that functions in statistics are often piece-wise constant or only piece-wise continuous, while, for example, in analysis continuous or even differentiable functions are the rule rather than the exception. When it comes to displaying data, the situation changes drastically. Usually, statistical data are discrete. Smoothness properties that simplify display of analytical data are not available for statistical data. So visualisations adapted to the needs of statistics are required.

4.2 Dimensions

If we move from one dimension to higher dimensions, there are challenges for theoretical considerations as well as for graphical displays. Again, linear models can serve as leading or warning examples.

The challenge may result from serious problems. Even under regularity conditions, the distribution on higher-dimensional spaces can be complex beyond limits. The classical identification problems for functions and spaces from analysis and geometry give a foretaste of what we have to cope with for the analysis of probability distributions.

Besides, there are home-made problems that are generated by choices made in the initial step.

An example of home-made problems can be illustrated for linear models. The interpretation of the Gauss-Markov estimator as a linear projection shows that seemingly only the coefficients for single regressors are estimated. In fact, a vector in the vector space spanned by the regressors is estimated; the attribution to individual regressors is mere linear algebra. The allocation to individual components does not depend on the influence of the individual regressor, but on the joint geometry of all the regressors. There is a direct interpretation of the coefficients only if the regressors form an orthogonal basis. For example if we duplicate the list of regressors in a linear model, the space does not change. The calculation in coordinates becomes more complicated because the regressors do not form a basis now, but from an abstract point of view, the information is unchanged. However, if there is no exact duplicate, but minute deviations (by minimal "error", rounding, transformations), the situation changes drastically. With respect to the Gauss-Markov estimator, only the space spanned by the regressors is relevant, and even minimal changes in the duplicate copy can duplicate the dimension of this space. This is an example of a home-made problem.

This and other examples are reason enough to analyse the relation between the variables more precisely. In the case of regression, for example, this not only refers to the relation between response and regressors, but, as illustrated by the recent example, also to the relations between the regressors.

To keep the relation to the regression problem and resort to experiences in this area, we use a formal imbedding of the regression problem in a more general framework. In regression, we had a singled-out variable, the response. The distribution of this variable was to be modelled as a function of the remaining variables, the regressors X. Now we combine response and regressor to a joined data vector $Z = (X, Y)$ and will discuss the joined distribution of Z. We retrieve the regression problem in this more general framework: for the regression problem we were looking for an estimator for the mean value function m in the model

$$Y = m(X) + \varepsilon.$$

In the general framework we consider the joint distribution of X and Y. The regression model becomes

$$Y = E(Y|X) + \varepsilon$$

and for now we have the identification $m(X) = E(Y|X)$.

If we are indeed interested in the regression model, additional work has to be done. An estimation of the conditional expected value $E(Y|X)$ is not the same as the estimation of a regression function $m(X)$. In the regression problem we have made no assumptions about the distribution of X. To infer from $E(Y|X)$ (or a estimator of this) to $m(X)$ we must establish that the estimation is independent of the distribution assumptions about X. For present purposes the distinction is irrelevant. We can allow ourselves a temporary ignorance.

So the general framework of this chapter is that we analyse data $(Z_i)_{i=1,\ldots,n}$, where the single observations take values in $\mathbb{R}^q$. The regression model is contained in this setting as $Z_i = (X_i, Y_i)$.

If we have essentially linear structures, we often can analyse higher-dimensional structures with methods that have been developed for one-dimensional models. We may have to modify these methods, or use them stepwise taking residuals. But they may help us to detect the essential structures. They fail, however, if higher dimensionality combines with non-linearity. In that case, more specific methods are needed.

4.3 Selections

Originally a selection indicated a selection of observations. For the graphical display, a selection is associated with a specification of display attributes (such as colour, plot character, line width). All variable values that belong to observations in the selection are displayed with these attributes. This allows us to follow links by selection in various plots ("***linking***"). With linking, selections can help us detect structures in linked plots.

In practical data analysis, selections are varied dynamically ("***brushing***"), to group corresponding observations. Brushing with linked plots is an important tool in interactive data analysis. In statistical terms, selections are used for model selection. They correspond to local models: instead of using a global, possibly rather complex for the data, for each selection a (hopefully simpler) model is used which only applies to the data in this selection.

Unfortunately, linking is not supported directly by R. We ourselves have to ensure that selections are displayed using their accordant attributes. Moreover, the syntactic representation of selections is not unified. In function calls, selections can be represented as a `selection` parameter, or by a *`group`* variable, or as a condition in a formula expression. So in each case we have to resort to ad-hoc solutions.

R is restricted mainly to static selections. Brushing is only possible in a rudimentary form.

Selections are illustrated in context in the following sections.

4.4 Projections

As a first example we look at a data set published in a comparison of different kind of diabetes ([37]). The data set is provided, for example, in *`library(locfit)`* [26] as *`data(chemdiab)`*. The variables refer to laboratory values on glucose metabolism (see Table 4.6). We get access to the datta a first summary with

Input

```
data(chemdiab)
summary(chemdiab)
```

As in the original publication we omit the relative weight from our discussion. The chemical classification *`cc`* is derived from the metabolic data. So it does not contain any information on its own. For orientation, we use it as a classification marker, that is, we use the selections *`cc = Chemical_Diabetic, Normal, Overt_Diabetic`*. Essentially, the data set is four-dimensional with the variables *`fpg, ga, ina, sspg`*.

4.4.1 Marginal Distributions and Scatter Plot Matrices

We can try to analyse multi-dimensional distributions by looking at the two-dimensional ***marginal distributions*** for all pairs of variables. The corresponding graphical display is called a ***scatterplot matrix***. In R, it is implemented as function `pairs()`.

Name	Variable	Unit, Remarks
rw	relative body weight	
fpg	plasma glucose (after overnight fast)	[mg/100 ml]
ga	glucose level integrated over about 3 hours tolerance test	[mg/100 ml × h]
ina	insulin level integrated over about 3 hours tolerance test	[μU/100 ml × h]
sspg	plasma glucose (steady state)	[mg/100 ml]
cc	classification	chemical, normal, open diabetes

Table 4.6 *Diabetes data set: variables*

---

help(pairs)

`pairs` *Scatterplot Matrices*

Description

A matrix of scatterplots is produced.

Usage

```
pairs(x, ...)

## S3 method for class 'formula':
pairs(formula, data = NULL, ..., subset,
      na.action = stats::na.pass)

## Default S3 method:
pairs(x, labels, panel = points, ...,
      lower.panel = panel, upper.panel = panel,
      diag.panel = NULL, text.panel = textPanel,
      label.pos = 0.5 + has.diag/3,
      cex.labels = NULL, font.labels = 1,
      row1attop = TRUE, gap = 1)
```

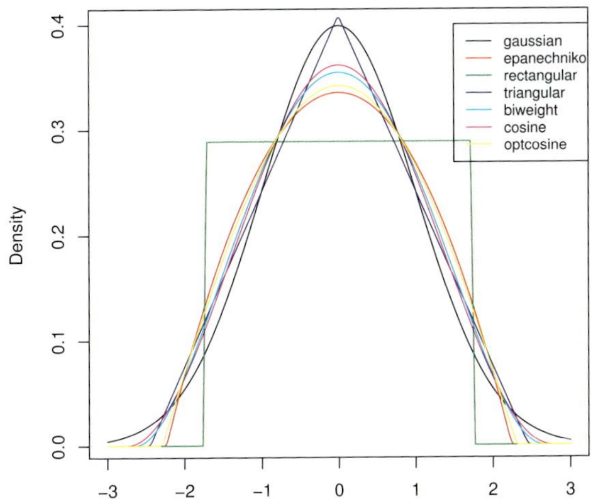

Colour Figure 1: *Kernels in* R. *See Example 1.4 (page 11) and Table 1.9 (page 12).*

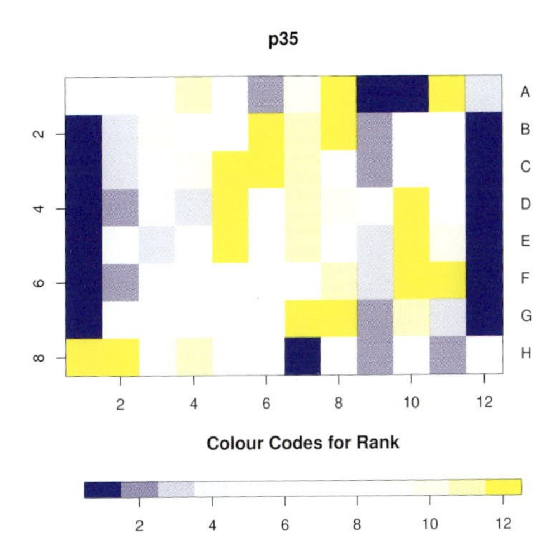

Colour Figure 2: *Titre plate diagnostics. See Example 2.12 (page 93) and Example 2.13 (page 95).*

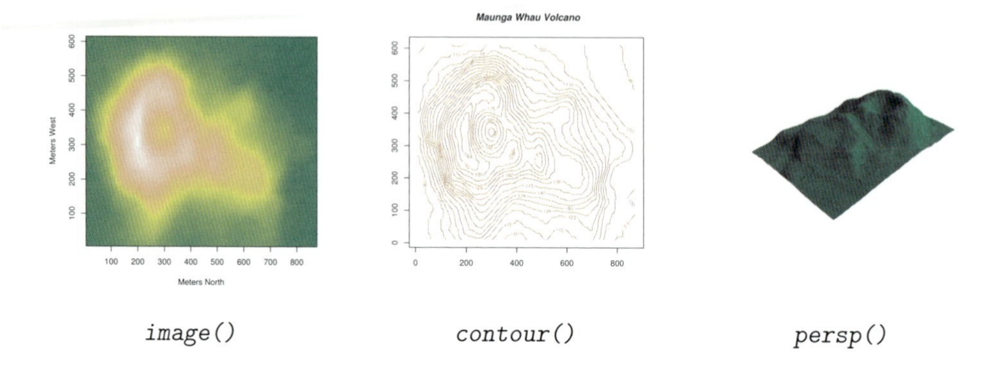

Colour Figure 3: *3d Surface displays using base graphics. See Example 4.1 (page 140).*

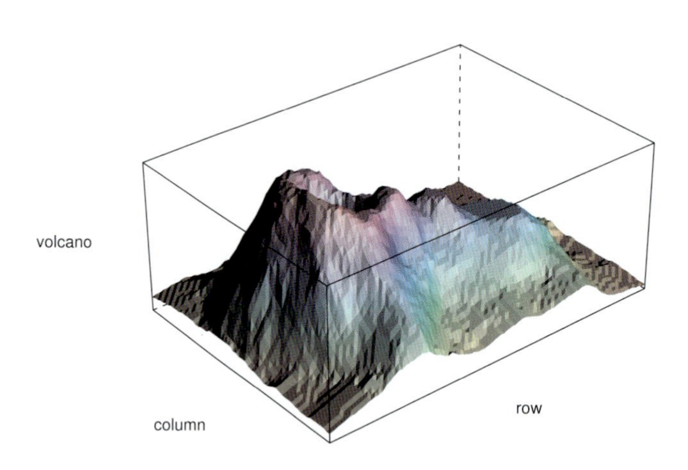

Colour Figure 4: *3d Surface display using lattice graphics. See Example 4.2 (page 142).*

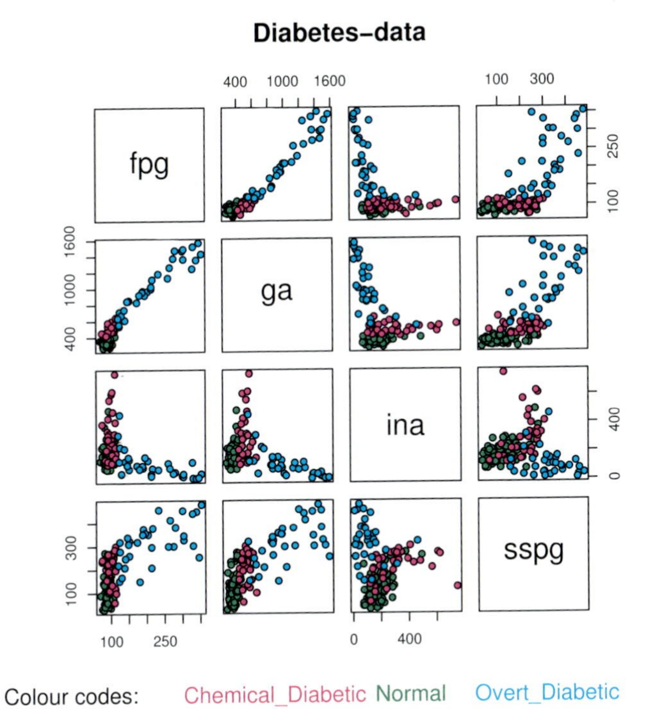

Colour Figure 5: *Diabetes data scatterplot matrix. See Section 4.4.1 (page 145).*

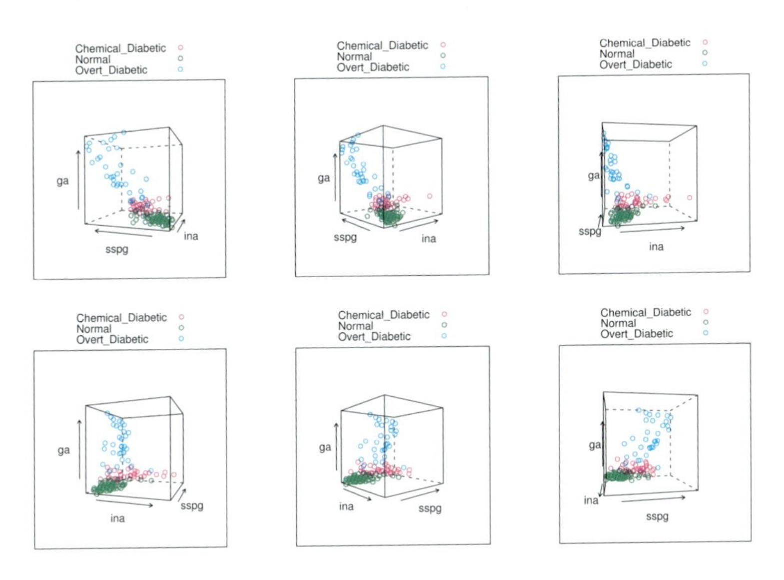

Colour Figure 6: *Diabetes data rotation. See Section 4.4.1 (page 145).*

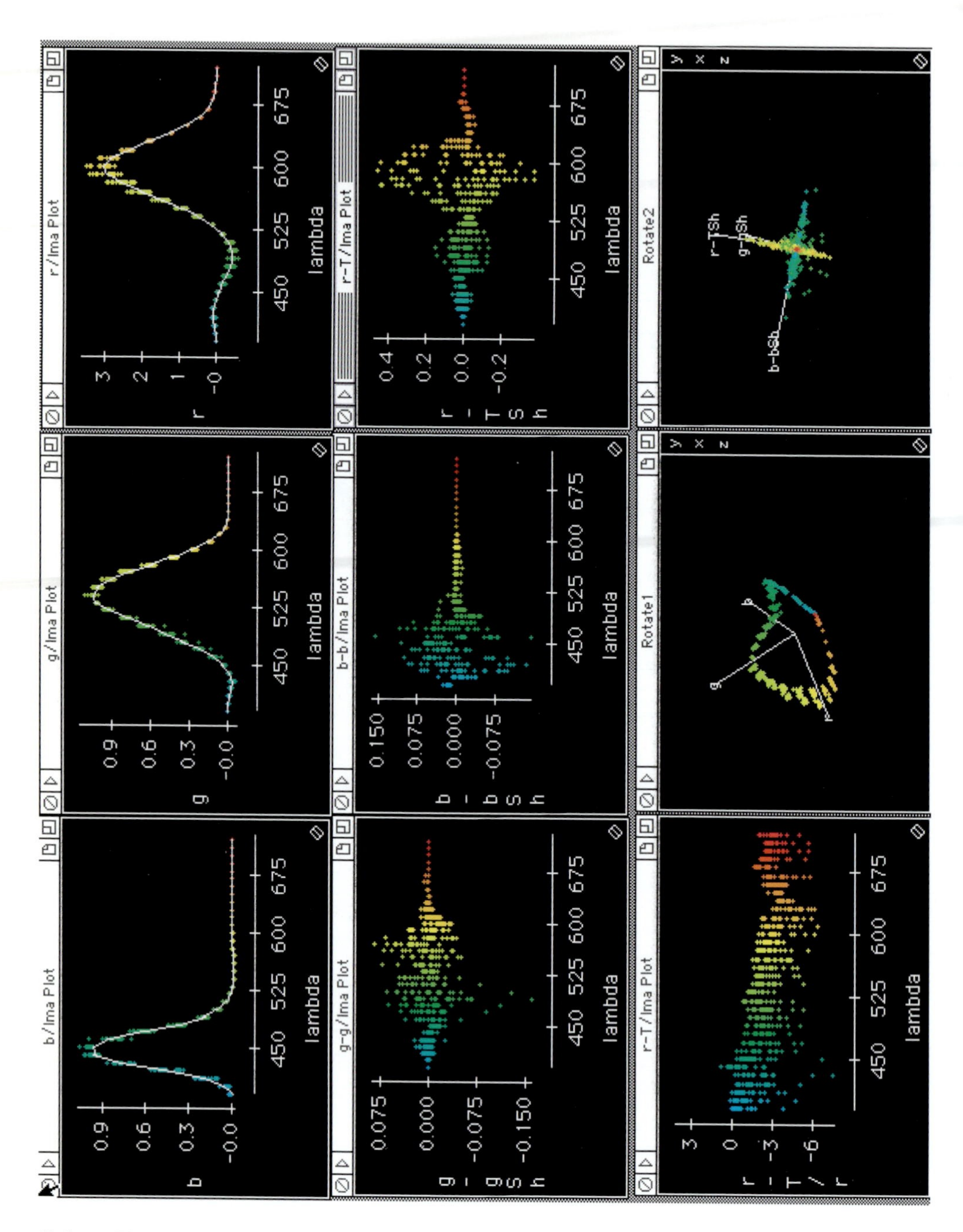

Colour Figure 7: *Colour perception data. Displays prepared with Data Desk [50]. See Section 4.4.3 (page 153). For an animated display and background information, see* <*http://www.statlab.uni-heidelberg.de/data/color/*>.

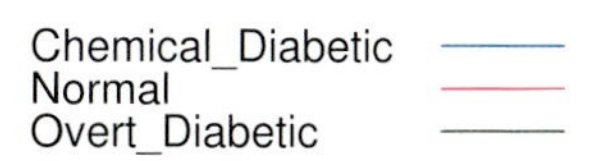

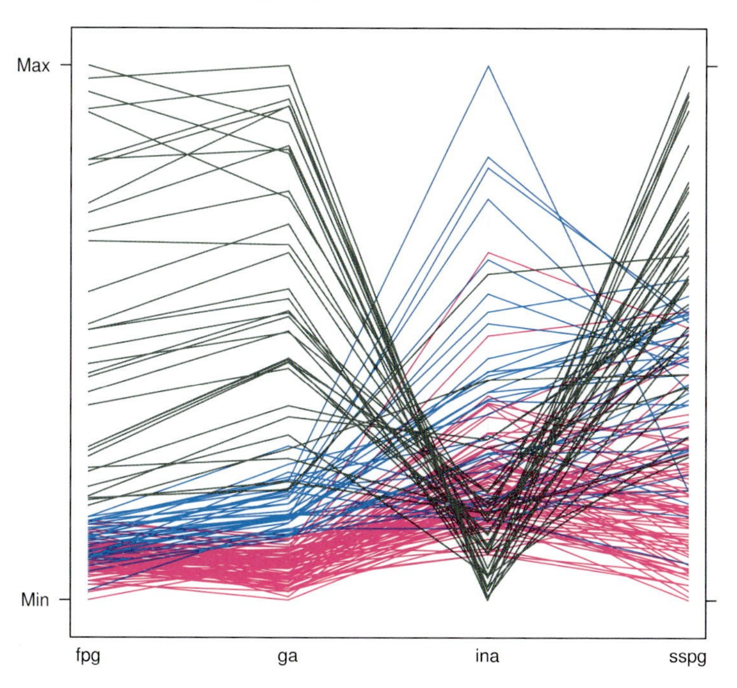

Colour Figure 8: *Diabetes data. See Section 4.4.4 (page 155).*

Colour Figure 9: *Iris species. See Section 4.6 (page 162).*

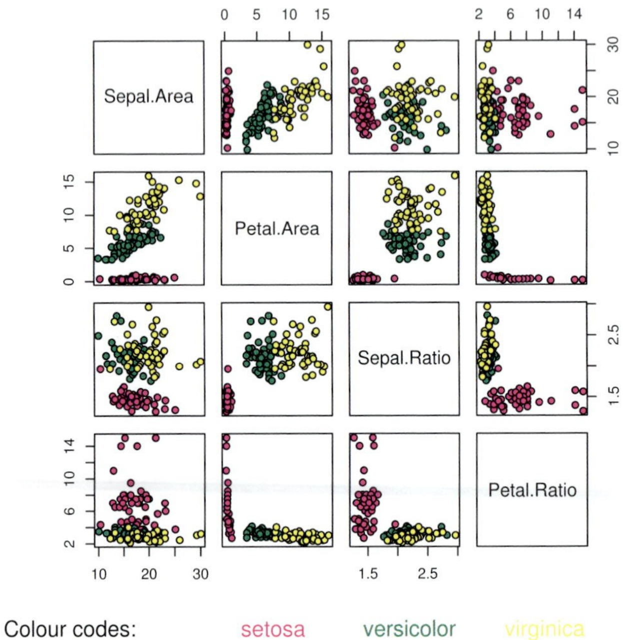

Colour Figure 10: *Iris species. See Section 4.6 (page 162).*

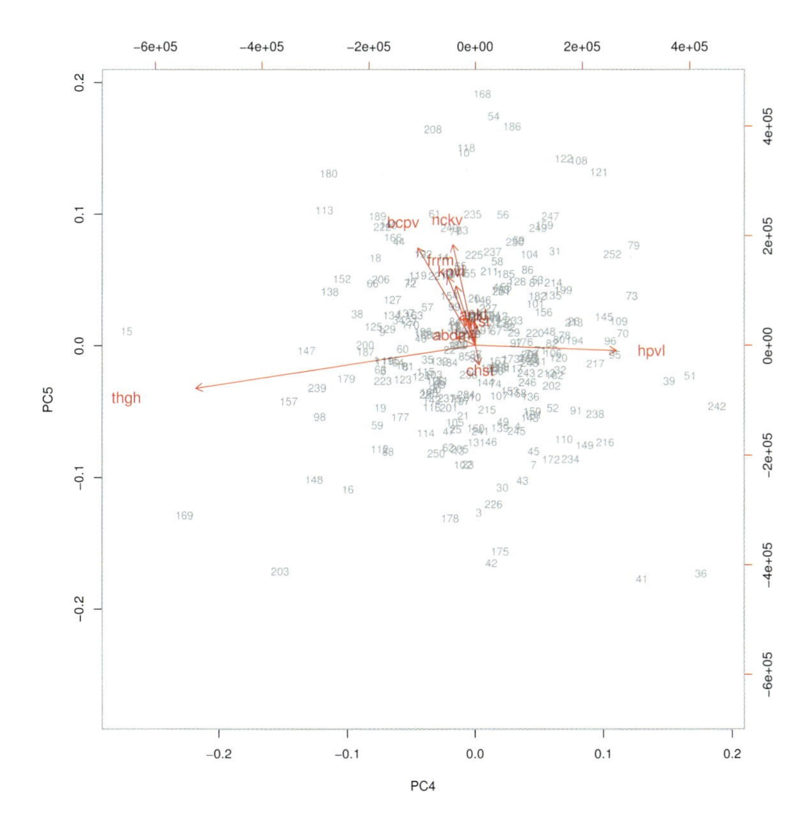

Colour Figure 11: *Biplot for principal components 4 and 5 of the fat data set. See Example 4.5 (page 186).*

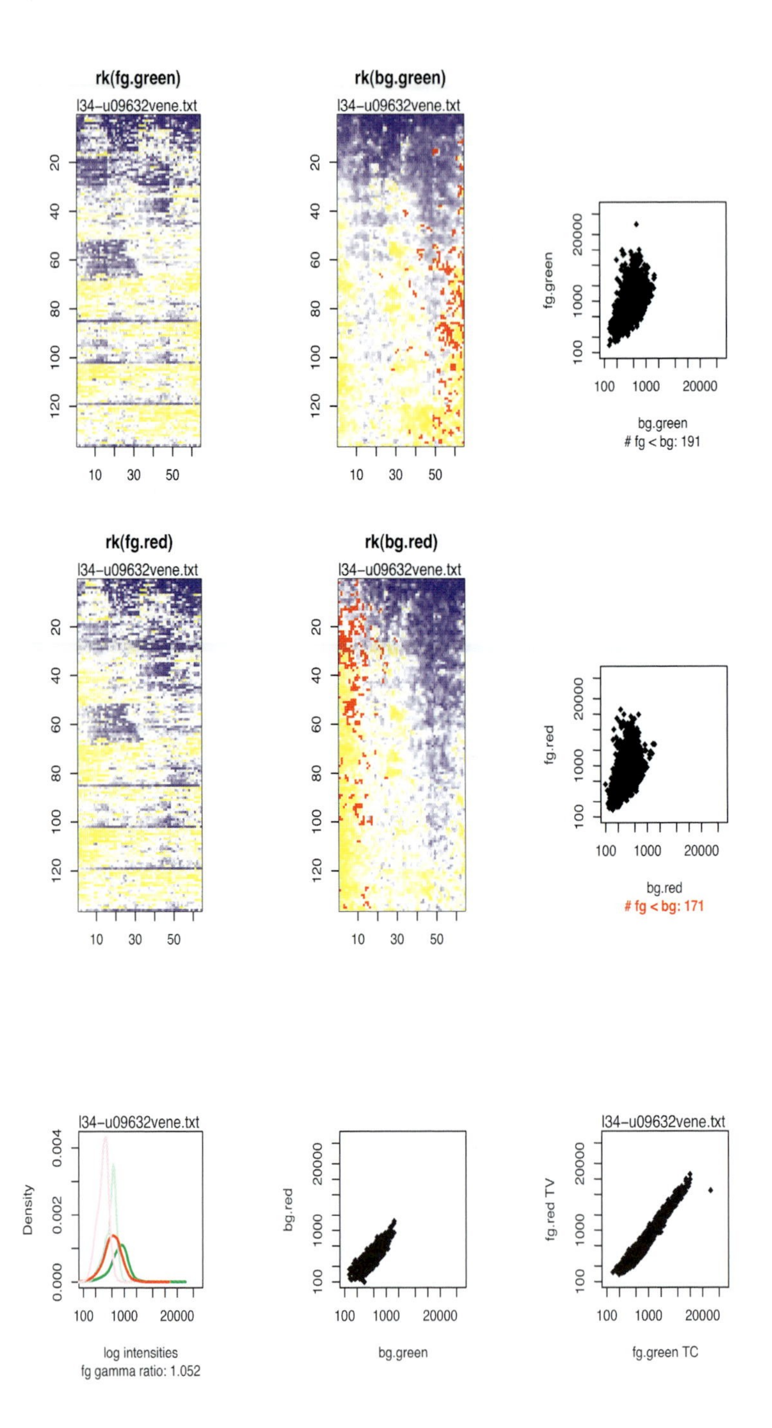

Colour Figure 12: *A single* 4227×4 *dimensional observation from a micro array experiment. See Section 4.8 (page 189).*

Arguments

`x`	the coordinates of points given as numeric columns of a matrix or dataframe. Logical and factor columns are converted to numeric in the same way that `data.matrix` does.
`formula`	a formula, such as `~ x + y + z`. Each term will give a separate variable in the pairs plot, so terms should be numeric vectors. (A response will be interpreted as another variable, but not treated specially, so it is confusing to use one.)
`data`	a data.frame (or list) from which the variables in `formula` should be taken.
`subset`	an optional vector specifying a subset of observations to be used for plotting.
`na.action`	a function which indicates what should happen when the data contain `NA`s. The default is to pass missing values on to the panel functions, but `na.action = na.omit` will cause cases with missing values in any of the variables to be omitted entirely.
`labels`	the names of the variables.
`panel`	`function(x,y,...)` which is used to plot the contents of each panel of the display.
`...`	arguments to be passed to or from methods. Also, graphical parameters can be given as can arguments to `plot` such as `main`. `par("oma")` will be set appropriately unless specified.
`lower.panel, upper.panel`	separate panel functions to be used below and above the diagonal respectively.
`diag.panel`	optional `function(x, ...)` to be applied on the diagonals.
`text.panel`	optional `function(x, y, labels, cex, font, ...)` to be applied on the diagonals.
`label.pos`	`y` position of labels in the text panel.
`cex.labels, font.labels`	graphics parameters for the text panel.
`row1attop`	logical. Should the layout be matrix-like with row 1 at the top, or graph-like with row 1 at the bottom?
`gap`	Distance between subplots, in margin lines.

Details

The *ij*th scatterplot contains `x[,i]` plotted against `x[,j]`. The scatterplot can be customised by setting panel functions to appear as something completely different. The off-diagonal panel functions are passed the appropriate columns of `x` as `x` and `y`: the diagonal panel function (if any) is passed a single column, and the `text.panel` function is passed a single `(x, y)` location and the column name.

The graphical parameters `pch` and `col` can be used to specify a vector of plotting symbols and colors to be used in the plots.

The graphical parameter `oma` will be set by `pairs.default` unless supplied as an argument.

A panel function should not attempt to start a new plot, but just plot within a given coordinate system: thus `plot` and `boxplot` are not panel functions.

By default, missing values are passed to the panel functions and will often be ignored within a panel. However, for the formula method and `na.action = na.omit`, all cases which contain a missing values for any of the variables are omitted completely (including when the scales are selected).

Author(s)

Enhancements for R 1.0.0 contributed by Dr. Jens Oehlschlaegel-Akiyoshi and R-core members.

References

Becker, R. A., Chambers, J. M. and Wilks, A. R. (1988) *The New S Language.* Wadsworth & Brooks/Cole.

Examples

```
pairs(iris[1:4], main = "Anderson's Iris Data -- 3 species",
      pch = 21, bg = c("red", "green3", "blue")[unclass(iris$Species)])

## formula method
pairs(~ Fertility + Education + Catholic, data = swiss,
      subset = Education < 20, main = "Swiss data, Education < 20")

pairs(USJudgeRatings)

## put histograms on the diagonal
panel.hist <- function(x, ...)
{
    usr <- par("usr"); on.exit(par(usr))
    par(usr = c(usr[1:2], 0, 1.5) )
    h <- hist(x, plot = FALSE)
    breaks <- h$breaks; nB <- length(breaks)
    y <- h$counts; y <- y/max(y)
    rect(breaks[-nB], 0, breaks[-1], y, col="cyan", ...)
}
pairs(USJudgeRatings[1:5], panel=panel.smooth,
      cex = 1.5, pch = 24, bg="light blue",
      diag.panel=panel.hist, cex.labels = 2, font.labels=2)

## put (absolute) correlations on the upper panels,
## with size proportional to the correlations.
panel.cor <- function(x, y, digits=2, prefix="", cex.cor)
{
    usr <- par("usr"); on.exit(par(usr))
    par(usr = c(0, 1, 0, 1))
    r <- abs(cor(x, y))
```

```
        txt <- format(c(r, 0.123456789), digits=digits)[1]
        txt <- paste(prefix, txt, sep="")
        if(missing(cex.cor)) cex.cor <- 0.8/strwidth(txt)
        text(0.5, 0.5, txt, cex = cex.cor * r)
    }
    pairs(USJudgeRatings, lower.panel=panel.smooth, upper.panel=panel.cor)
```

We use the chemical diabetes classes `cc` to define selections. We assign a colour to each of these selections. This is the linking attribute that allows us to trace relations between the plots. To provide a documentation of this linking, we have to control the graphics and modify the plots. Using the parameter `oma` we generate an outer margin to give space for a caption.

***Example* 4.3: Scatterplot Matrix for Diabetes Data**

Input

```
pairs(~fpg + ga + ina + sspg, data = chemdiab,     pch = 21,
    main = "Diabetes-data",
     bg = c("magenta", "green3", "cyan")[unclass(chemdiab$cc)],
    oma = c(8, 8, 8, 8))
mtext(c("Colour codes:", levels(chemdiab$cc)),
    col = c("black", "magenta", "green3", "cyan"),
    at = c(0.1, 0.4, 0.6, 0.8), side = 1, line = 2)
```

Output: see Figure 4.1.

The `pairs()` function only controls the "layout" of the matrix, the selection and placement of the projections. The display in the individual plot tiles can be controlled in the function call. The defaults give the names of the variables in the diagonal tiles, and pair-wise scatterplots in the off-diagonal tiles.

Exercise 4.1	
	Generate a scatterplot matrix for the diabetes data set that shows a histogram of the variables in the diagonal tiles. *Hint:* See `help(pairs)`.

Certain aspects of the distribution can be read easily from the marginal distributions. Other geometrical structures are barely or not at all recognisable from the marginal distribution.

For example, there is an apparent linear relation between the glucose level after a fast `fpg` and the integrated glucose level under tolerance test. This can be seen from the two-dimensional marginal distributions, and it can be analysed using the methods for linear models.

This apparent relation propagates to the relations to the other variables `ina, sspg`. For this reason, `fpg` is not considered in detail in the original paper. What remains to be analysed

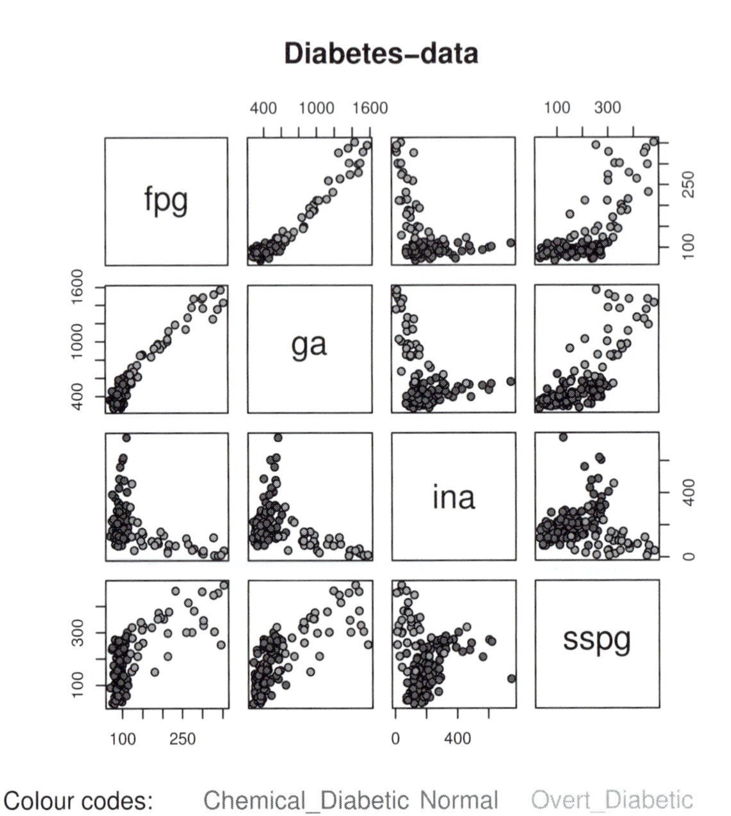

Figure 4.1 *Scatterplot matrix for diabetes data. See Colour Figure 5.*

are the variables `ga, ina, sspg`. The three-dimensional structure of this part of the data set is not easily detected from the marginal distributions. The data structure shows two lobes in three-dimensional space. Each lobe is essentially two-dimensional, but is imbedded in three-dimensional space like a floppy ear.

4.4.2 Projection Pursuit

Geometrical relations or stochastic dependencies that are not aligned along coordinate axes are in general not covered by marginal distributions. We can generalise the idea. Instead of two-dimensional marginal distributions, we can use arbitrary projections. We use `library(lattice)`, which supports a camera oriented view on 3d data.

The `grid` graphics and the `lattice` package provide support for multivariate representations. `grid` forms the basis. The original R graphics system implements a model that takes pen and paper as a model. A graphic port (paper) is opened, and lines, points or symbols are drawn. *grid* is a second graphics system, which follows a object/viewport model. Graphical objects in various locations and orientations are mapped in a visual space. `lattice` builds upon *grid*. <http://cm.bell-labs.com/cm/ms/departments/sia/project/trellis/> gives the underlying ideas for visualisation of multidimensional data that are implemented in `lattice`.

The graphics thus generated is output with *print()*. The parameter *split* allows dividing up the output area. Unfortunately, linking is broken here: *cloud()* can generate a caption, but this shows the colour scale that was in effect when the graphics started, not the colour scale effectively used in the output. Once again we have to interfere with the system, this time to adjust the colour table.

Input

```
library("lattice")

diabcloud <- function(y, where, more = TRUE, ...) {
    print(cloud(ga ~ ina + sspg, data = chemdiab, groups = cc,
        screen = list(x = -90, y = y), distance = .4, zoom = .6,
        auto.key = TRUE, ...),
        split = c(where, 3, 2), more = more)
}

supsym <- trellis.par.get("superpose.symbol")
supsymold <- supsym

supsym$col = c("magenta", "green3", "cyan")

trellis.par.set("superpose.symbol" = supsym)

diabcloud(y = 70, where = c(1, 1))
diabcloud(y = 40, where = c(2, 1))
diabcloud(y = 10, where = c(3, 1))
diabcloud(y = -20, where = c(1, 2))
diabcloud(y = -50, where = c(2, 2))
diabcloud(y = -80, where = c(3, 2), more = FALSE)

trellis.par.set("superpose.symbol" = supsymold)
rm(diabcloud, supsymold, supsym)
```

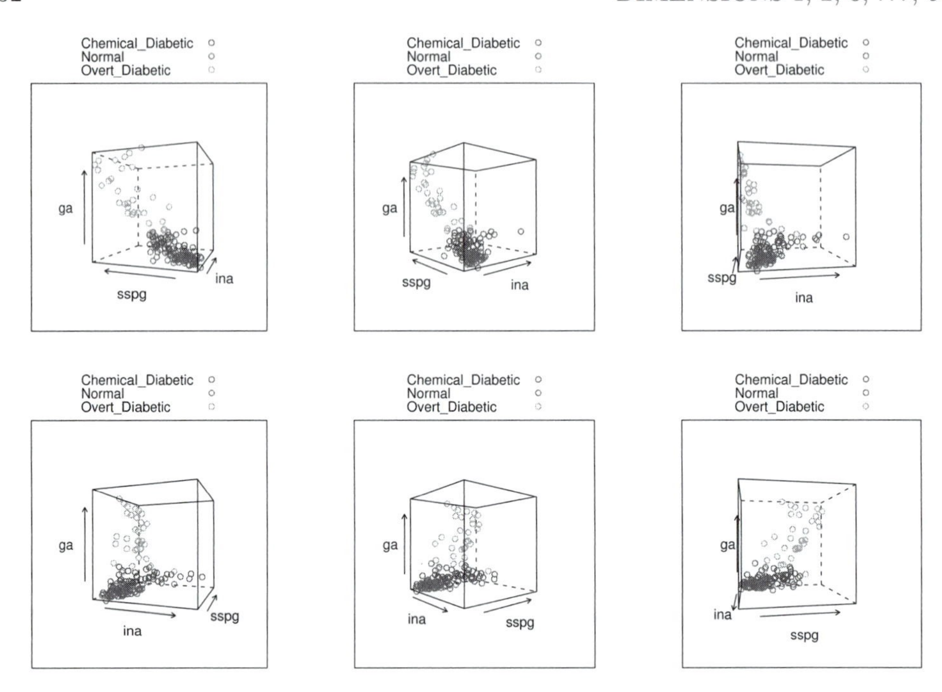

See Colour Figure 6.

Exercise 4.2	
	Modify this example so that you get an impression of the three-dimensional structure. Try to use an animated sequence. You can use *sys.wait()* if it is available on your system to control the time sequence, or use *devAskNewPage()* to give interactive control for new images.
	What is the difference between open diabetes and chemical diabetes?
	How does the normal group compare to both diabetes groups?

Even with a series of projections it is often not simple to identify a three-dimensional structure. Animation can help. There is support for animation in `library(rggobi)` [23]. This, however, requires ggobi, accessible in `<http://www.ggobi.org/>`, as additional software. If `library(rgl)` is installed, you get an interactive 3d plot with

```
points3d(chemdiab$ga, chemdiab$ina, chemdiab$sspg)}.
```

What has been done here ad hoc can be done systematically and generalised for arbitrary dimensions: for a data set in $\mathbb{R}^q$ one looks for "interesting" projections. For this, an index is defined that should indicate how interesting a projection is. Then a search is started to maximise this index. A family of statistical methods based on this idea can be found under the keyword ***projection pursuit***. `ggobi` contains implementations of projection pursuit for several indices. These can be accessed using the R functions in `library(rggobi)`.

4.4.3 Projections for Dimensions 1, 2, 3, ... 7

Projection methods try to identify structures of lower dimension in a higher-dimensional data set. The dimensions that can be identified are limited. If we project a structure with a dimension larger than the projection target, a typical projection covers the target space and no longer gives information.

How many dimensions can we capture? The graphical display in the plane gives, as lower bound, two dimensions. We can represent two-dimensional structures directly using Cartesian coordinates in the xy plane. (At least locally. The global structure might require additional dimensions.) Perception can reconstruct three-dimensional structures with the help of visual depth cues (such as a shade) or from sequences of 2d images. Animation can give the impression of changing 3d scenes and we reach four dimensions.

Additional information channels, as for example colour encoding, can elevate the dimension slightly, but effectively we stay with four to seven dimensions for a display.

Combination of several displays helps little beyond this limit. If we combine several displays, for example generalising the idea of a scatterplot matrix, we lose the capability of generating more complex structures by our perception. Instead, we have to work out the more complex structures by active comparison. The ability to perform simultaneous comparison of structures is limited, as is the number of displays that can be presented simultaneously on a page medium such as a screen or paper.

Four dimensions is easy. Going beyond seven is hard.

To illustrate an example in seven dimensions, we use a classical experiment on colour perception [47]. An observer (usually a person well trained in working with colours) is confronted with a target colour, and control over three light sources (red, green and blue). The task is to adjust the controllable light sources to match the target colour. The classical experiment was performed with 34 target colours varying in wavelength in steps of 10nm, with 20 observers.

To start with, we can view a single observation as a four-dimensional datum (target colour, intensities for red, green, blue). We can take the three marginals of the target colour by any of the intensities. This gives us the colour matching curves that are found in many textbooks on colour perception. (See Figure 4.2 top row.) We use colour encoding by target colour here as an additional cue, linking the windows. Several notes are necessary. First, the classical experiments have been made with a black background, at that time. Second, the colour cues are a rough approximation, not the exact target colour. Third, not all colours can be mixed from the primaries in this experiment. The observer had the possibility to set the control to negative values which would have the effect to change the target colour by blending. Fourth, as in all real data, there are measurement errors. One observation was off. We corrected it for our display.

The basic data set has four-dimensional observations. The target colour is already encoded as colour cue. We can combine the three response dimensions in a 3d scatterplot. An example is in Figure 4.2, bottom row middle. It tells us that essentially we have a one-dimensional data structure, imbedded non-linearly in 3d space.

From our experience with linear models we know that the data scatterplot only shows part of the information. The other information is contained in the residuals. Taking residuals from one-dimensional non-linear regression for the three components (Figure 4.2, middle row) gives us another three-dimensional view on the information contained in the data, combined in Figure 4.2, bottom row right. Using a (non-marginal) projection from this 3d structure tells us that again we have a non-linear structure: there is a disk, essentially in the plane characterised

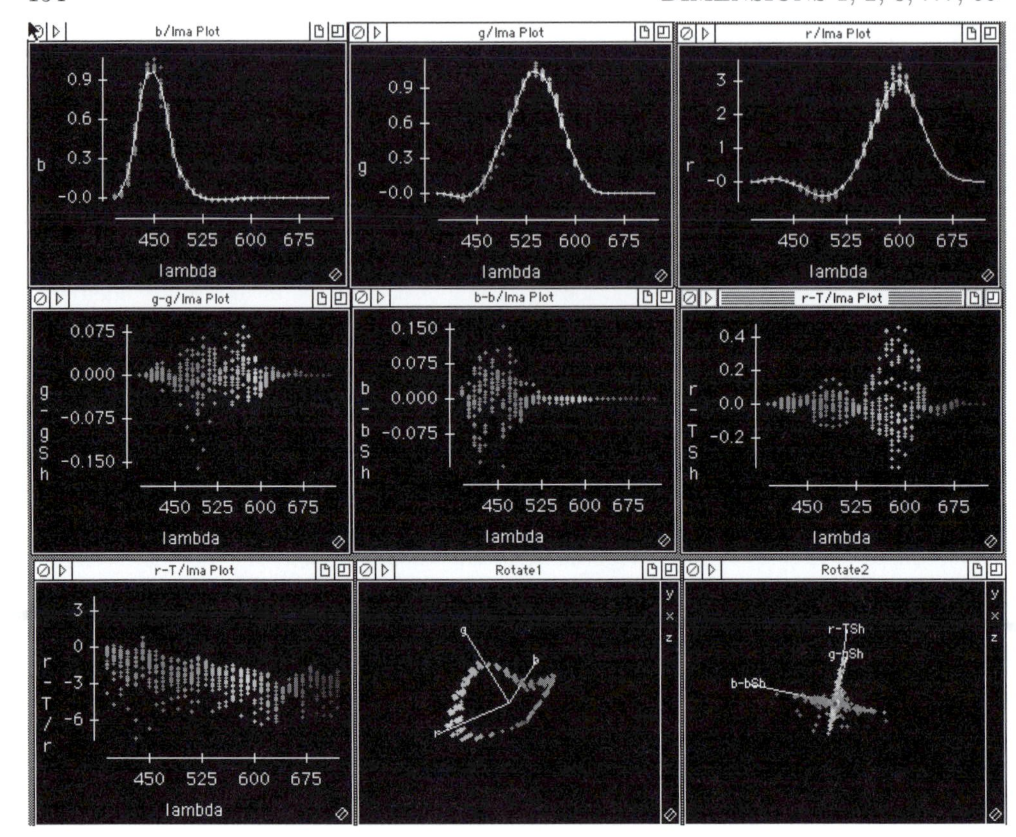

Figure 4.2 *Colour perception data. Displays prepared with Data Desk [50]. For an animated display and background information, see* `<http://www.statlab.uni-heidelberg.de/data/color/>`. *See Colour Figure 7.*

by red/green colours, and a rod-like structure orthogonal to this, essentially characterised by blue/yellow colours. So the complete data structure has seven dimensions, with one dimension (the target colour) used for linking and represented by colour.

We only used basic data analysis approaches to reveal this structure. Work on colour perception has proceeded. The conception nowadays is that there are essentially two systems in early colour perception, one combining the red/green channels, the other one based on the blue channel. We did not use this structural information. But the information we get from the residuals just fits.

4.4.4 Parallel Coordinates

The graphical display (in Cartesian coordinates) is initially restricted to one- and two-dimensional projections. But even for representation in a plane, the restriction to two dimensions is not predetermined, but is a consequence of our choice to display the data in Cartesian coordinates. Plot matrices break this dimension barrier by combining Cartesian coordinate systems.

Parallel coordinates use a parallel orientation of axes, not an orthogonal orientation. For frequencies with categorical variables this is a common display: (possibly stacked) bar charts use parallel coordinates. Function `parallel()` in `library(lattice)` implements parallel coordinates for quantitative variables as well. The marks on these axes that correspond to one case are joined by a zigzag line. This form of parallel coordinates was suggested by A. Inselberg ([18]). For a survey, see ([17]).

Input

```
library("lattice")
print(parallel(chemdiab[2:5], groups = chemdiab$cc,
    horizontal=FALSE,
    auto.key=TRUE))
```

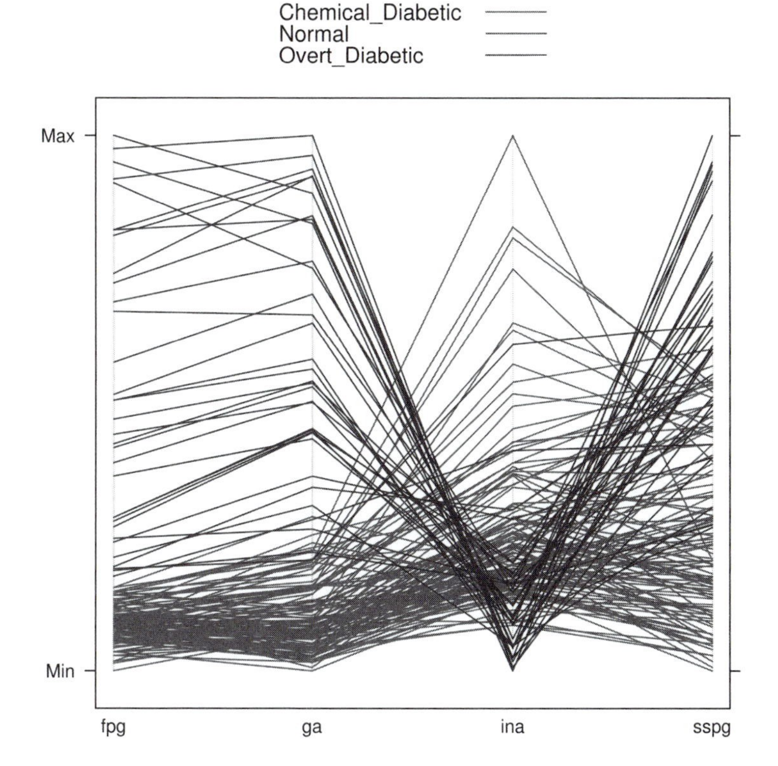

See Colour Figure 8.

The information is the same as that in previous plots of this data set. Using a different display gives a new look to the structure in this data set.

What is obvious at a first glance is that the insulin level `ina` behaves quite differently from the other variables. Parallel coordinate plots can help us to see this kind of particularity. Going beyond this, there is a certain duality between cartesian and parallel coordinates. For example, linear dependency in cartesian coordinates appears as a pencil in parallel coordinates.

To identify this kind of relation in parallel coordinates requires some training, as provided in [17].

Exercise 4.3	
	For the ***chemdiab*** data set, prepare a (written!) report about the relation between the variables that you can recognise in the parallel coordinate plot.
	Instead of using *chemdiab[2:5]* you can specify the variables explicitly as *chemdiab[c(2, 3, 4, 5)]*. This gives you control over the order of the variables. Compare two different sequences of the variables and note (in writing!) your observations. Which sequence of variables gives the simpler display? Which relations between the variables are visible in both? Which relations appear only in one of the arrangements?

4.5 Sections, Conditional Distributions and Coplots

From an abstract point of view, sections are conditional distributions of the type $P(\cdot \mid X = x)$. But they are only reliable where the section defines a condition that has positive probability. To make the idea of restricting the view on conditional distributions applicable to data, we thicken the sections. Instead of considering conditional distributions of the type $P(\cdot \mid X = x)$ we consider $P(\cdot \mid \|X - x\| < \varepsilon)$, where ε possibly can vary with x. In graphical representations of data this requires a series of plots showing only the part of the data set specified by the condition.

Statistically, projections lead to marginal distributions and sections to conditional distributions. In a certain sense, sections and projections are complementary: projections show structural features of low dimension. Sections are helpful to detect structural features of low codimension. For data analysis, both can be combined. The interplay of projections and sections is discussed in [10]. Like the dimension boundaries for projections there are boundaries for the codimension when using sections. We can only catch structures of small codimension. If the codimension is too large, a typical section is empty, hence it has no information.

As a first tool, R provides the possibility to analyse two variables ***conditioned*** on one or more additional variables. As a graphical display *coplot()* serves for this purpose. It is a variant of the plot matrix and shows in each tile the scatterplot of two variables, given the condition.

The coplot can be inspected for patterns. If the variables shown are stochastically independent of the conditioning variables, all plot elements show the same shape. The variables shown and the conditioning variables can then be de-coupled.

If the general shape coincides, but location and size vary, this hints at a (not necessarily linear) shift/scale relation. Additive models or variants of these can be used to model the relation between the variables shown and conditioning variables.

If the shape changes with varying condition, a major dependency structure or interaction may apply that needs more precise modelling.

help(coplot)

coplot	*Conditioning Plots*

Description

This function produces two variants of the **conditioning plots discussed in the reference below.**

Usage

```
coplot(formula, data, given.values, panel = points, rows, columns,
       show.given = TRUE, col = par("fg"), pch = par("pch"),
       bar.bg = c(num = gray(0.8), fac = gray(0.95)),
       xlab = c(x.name, paste("Given :", a.name)),
       ylab = c(y.name, paste("Given :", b.name)),
       subscripts = FALSE,
       axlabels = function(f) abbreviate(levels(f)),
       number = 6, overlap = 0.5, xlim, ylim, ...)
co.intervals(x, number = 6, overlap = 0.5)
```

Arguments

`formula` a formula describing the form of conditioning plot. A formula of the form `y ~ x | a` indicates that plots of `y` versus `x` should be produced conditional on the variable `a`. A formula of the form `y ~ x| a * b` indicates that plots of `y` versus `x` should be produced conditional on the two variables `a` and `b`.

All three or four variables may be either numeric or factors. When `x` or `y` are factors, the result is almost as if `as.numeric()` was applied, whereas for factor `a` or `b`, the conditioning (and its graphics if `show.given` is true) are adapted.

`data` a data frame containing values for any variables in the formula. By default the environment where `coplot` was called from is used.

`given.values` a value or list of two values which determine how the conditioning on `a` and `b` is to take place.

When there is no `b` (i.e., conditioning only on `a`), usually this is a matrix with two columns each row of which gives an interval, to be conditioned on, but is can also be a single vector of numbers or a set of factor levels (if the variable being conditioned on is a factor). In this case (no `b`), the result of `co.intervals` can be used directly as `given.values` argument.

`panel` a `function(x, y, col, pch, ...)` which gives the action to be carried out in each panel of the display. The default is `points`.

`rows` the panels of the plot are laid out in a `rows` by `columns` array. `rows` gives the number of rows in the array.

`columns`	the number of columns in the panel layout array.
`show.given`	logical (possibly of length 2 for 2 conditioning variables): should conditioning plots be shown for the corresponding conditioning variables (default `TRUE`)
`col`	a vector of colors to be used to plot the points. If too short, the values are recycled.
`pch`	a vector of plotting symbols or characters. If too short, the values are recycled.
`bar.bg`	a named vector with components `"num"` and `"fac"` giving the background colors for the (shingle) bars, for numeric and factor conditioning variables respectively.
`xlab`	character; labels to use for the x axis and the first conditioning variable. If only one label is given, it is used for the x axis and the default label is used for the conditioning variable.
`ylab`	character; labels to use for the y axis and any second conditioning variable.
`subscripts`	logical: if true the panel function is given an additional (third) argument `subscripts` giving the subscripts of the data passed to that panel.
`axlabels`	function for creating axis (tick) labels when x or y are factors.
`number`	integer; the number of conditioning intervals, for a and b, possibly of length 2. It is only used if the corresponding conditioning variable is not a `factor`.
`overlap`	numeric < 1; the fraction of overlap of the conditioning variables, possibly of length 2 for x and y direction. When overlap < 0, there will be *gaps* between the data slices.
`xlim`	the range for the x axis.
`ylim`	the range for the y axis.
`...`	additional arguments to the panel function.
`x`	a numeric vector.

Details

In the case of a single conditioning variable `a`, when both `rows` and `columns` are unspecified, a 'close to square' layout is chosen with `columns >= rows`.

In the case of multiple `rows`, the *order* of the panel plots is from the bottom and from the left (corresponding to increasing `a`, typically).

A panel function should not attempt to start a new plot, but just plot within a given coordinate system: thus `plot` and `boxplot` are not panel functions.

The rendering of arguments `xlab` and `ylab` is not controlled by `par` arguments `cex.lab` and `font.lab` even though they are plotted by `mtext` rather than `title`.

Value

`co.intervals(., number, .)` returns a (`number` $\times$ 2) `matrix`, say `ci`, where `ci[k,]` is the `range` of `x` values for the `k`-th interval.

References

Chambers, J. M. (1992) *Data for models.* Chapter 3 of *Statistical Models in S* eds J. M. Chambers and T. J. Hastie, Wadsworth & Brooks/Cole.

Cleveland, W. S. (1993) *Visualizing Data.* New Jersey: Summit Press.

See Also

pairs, panel.smooth, points.

Examples

```
## Tonga Trench Earthquakes
coplot(lat ~ long | depth, data = quakes)
given.depth <- co.intervals(quakes$depth, number=4, overlap=.1)
coplot(lat ~ long | depth, data = quakes, given.v=given.depth, rows=1)

## Conditioning on 2 variables:
ll.dm <- lat ~ long | depth * mag
coplot(ll.dm, data = quakes)
coplot(ll.dm, data = quakes, number=c(4,7), show.given=c(TRUE,FALSE))
coplot(ll.dm, data = quakes, number=c(3,7),
       overlap=c(-.5,.1)) # negative overlap DROPS values

## given two factors
Index <- seq(length=nrow(warpbreaks)) # to get nicer default labels
coplot(breaks ~ Index | wool * tension, data = warpbreaks,
       show.given = 0:1)
coplot(breaks ~ Index | wool * tension, data = warpbreaks,
       col = "red", bg = "pink", pch = 21,
       bar.bg = c(fac = "light blue"))

## Example with empty panels:
with(data.frame(state.x77), {
coplot(Life.Exp ~ Income | Illiteracy * state.region, number = 3,
       panel = function(x, y, ...) panel.smooth(x, y, span = .8, ...))
## y ~ factor -- not really sensical, but 'show off':
coplot(Life.Exp ~ state.region | Income * state.division,
       panel = panel.smooth)
})
```

We illustrate coplots with the "Quakes" data set. This data set gives the geographical longitude and latitude of a series of earthquakes near the islands of Fiji, together with the depth of the seismic centre. We use the geographical longitude and latitude as variables for marginal projection, and the depth as the covariable defining the sections. For the x and y variables, orientation can be controlled by setting `xlim` resp. `ylim` appropriately. For the conditioning variable, we need a trick. We re-code the depth, so that in graphical representations large depth points downwards.

Input

```
quakes$depth <- -quakes$depth
given.depth <- co.intervals(quakes$depth, number = 4, overlap = .1)
coplot(lat ~ long | depth, data = quakes,
    given.values = given.depth,  columns = 1)
```

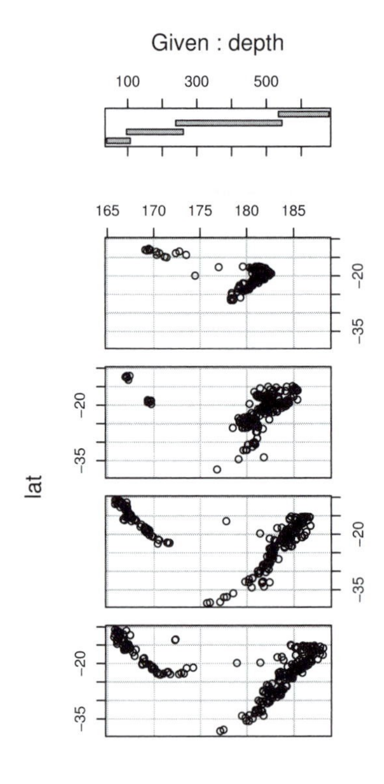

Analogous for two covariables, the depth and the magnitude of the earthquake.

Input

```
coplot(lat ~ long | mag * depth, data = quakes, number = c(5, 4))
```

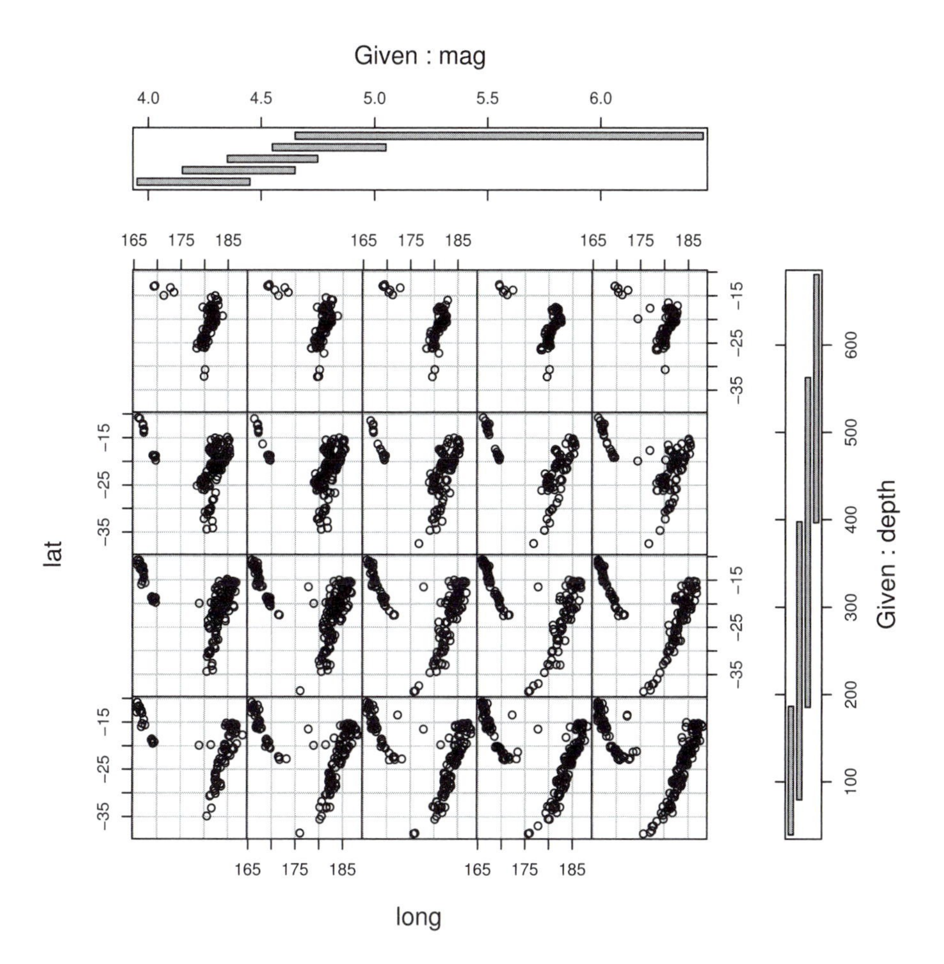

Exercise 4.4	Earthquakes
	Analyse the "quakes" data set. Summarise your results in a report. Try to specify a formal model.
	How is the geographic position related to the depth?
	Can you identify relations between depth and magnitude of the earthquake? (You may have to choose a different model formula in `coplot()`.)

The idea of the coplots is generalised in the trellis displays (see [6]). Trellis displays are implemented in R in *library("lattice")*.

Input

```
library("lattice")
Depth <- equal.count(quakes$depth, number = 4, overlap = .1)
print(xyplot(lat ~ long | Depth, data = quakes, columns = 1, layout = c(1, 4)))
```

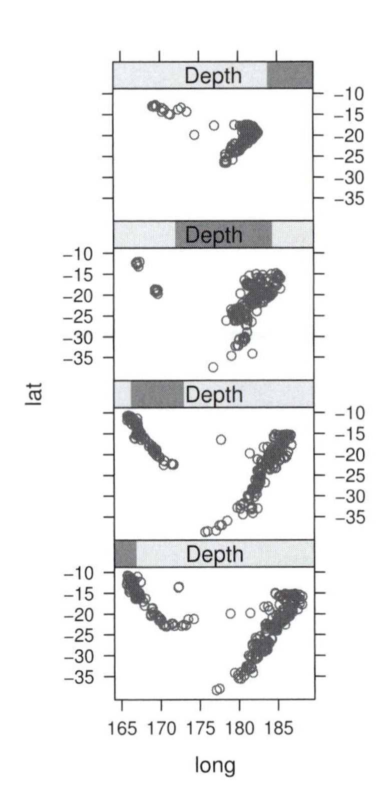

4.6 Transformations and Dimension Reduction

Variables often come in a scale that is suggested by the measurement process or by conventions of the trade. This does not necessarily correspond to what would be suggested by the subject matter or what is best used for statistical modelling. This scale has some arbitrariness:

- For an acoustic perception experiment, for example, the strength of the stimulus can be described by energy or by acoustic pressure [phone]. One scale is transformed to the other by a logarithm. The Weber-Fechner law in psychology says that for the human perception the (logarithmic) phone scale is appropriate. Response models or assumptions on errors become very complicated if you choose the other scale.

- Fuel consumption is reported in the USA in miles per gallon, in Europe as litres per 100 km. Up to conversion constants, one variable is the inverse of the other. Assumptions like normal distribution of errors can hold on one of the scales, but in no way on both. When we want to model fuel consumption, the description in litres per 100 km seems to lead to simpler statistical models in terms of weight, power and other parameters describing the car. Analysis in terms of miles per gallon can become rather complicated.

The choice of an appropriate scale for the variables can be a critical step in the analysis. Sometimes it is helpful to first introduce transformations and constructed variables, and then to reduce dimensions and to determine the effective variables in a second step.

Coordinate systems are not pre-set canonically. This already applies to univariate problems. For univariate problems, transformations of coordinate systems are relatively simple. Modelling of the error distribution on the one hand and transformation the data to a standard distribution on the other are in a certain way exchangeable. In higher dimension situations, suitable transformation families sometimes are not available or not accessible, and the structure of a problem may critically depend on the choice of suitable coordinates. A choice of the coordinate representation based on the subject matter often is preferable to an automatic selection.

We illustrate this with Anderson's "Iris" data set. The data set has five dimensions: four quantitative variables (length and width of petal sepal of iris blossoms) and a categorical variable (the species: *Iris setosa canadensis*, *Iris versicolor*, *iris virginica*).[1] The question is to find a classification of the species by the four quantitative variables.

Iris setosa | *Iris versicolor* | *Iris virginica*

Figure 4.3 *Iris species. See Colour Figure 9.*

The structure is similar to that of the diabetes data set `chemdiab`. The classification by `iris$Species`, however, is here a given (external) classification, in contrast to the derived classification `chemdiab$cc`. While in the case of `chemdiab` a general description was looked for, the aim is now to find a classification rule that classifies `iris$Species` by means of the other variables.

In this example, the species define the selection groups that are of interest.

To get a first impression we can view the four quantitative variables by species. The standard conventions of R make this cumbersome. Since species is a categorical variable, this causes R to switch from a scatterplot display to box-and-whisker plots when using `plot()`.

[1] Photos: *Iris setosa*: C. Hensler, The Rock Garden; *Iris versicolor* and *Iris virginica*: D. Kramb; The Species Iris Group of North America. With kind permission.

Input

```
oldpar <- par(mfrow = c(2, 2))
iriscols <- c("magenta", "green3", "yellow")

plot(iris$Species, iris$Petal.Length,
    ylab = '', main = 'Petal Length', col = iriscols)
plot(iris$Species, iris$Petal.Width,
    ylab = '', main = 'Petal Width', col = iriscols)
plot(iris$Species, iris$Sepal.Length,
    ylab = '', main = 'Sepal Length', col = iriscols)
plot(iris$Species, iris$Sepal.Width,
    ylab = '', main = 'Sepal Width', col = iriscols)

par(oldpar)
```

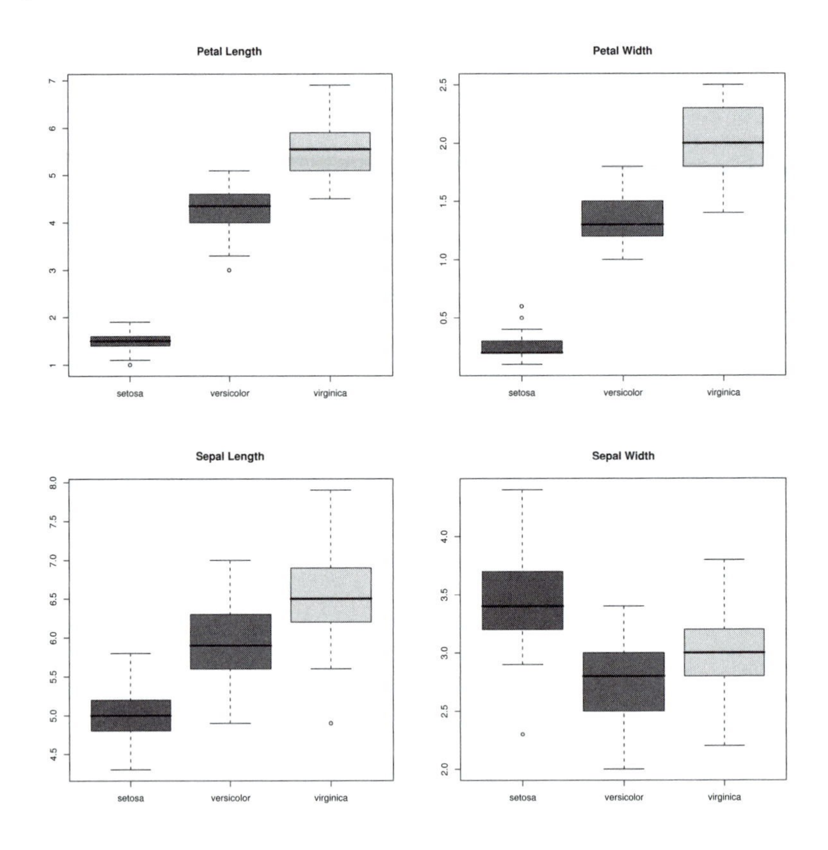

We could modify the R functions to show a scatterplot of the individual variables by group. Instead we resume to *grid* and *lattice* and use the function *stripplot()*. Since given the measurement precision in this experiment we have ties, that is the same value is recorded in multiple cases, we jitter the data to present them separately.

Input

```
library("lattice")

print(stripplot(Petal.Length ~ Species, data = iris,
    jitter = TRUE, ylab = '', main = 'Petal Length'),
    split = c(1, 1, 2, 2), more = TRUE)
print(stripplot(Petal.Width ~ Species, data = iris,
    jitter = TRUE, ylab = '', main = 'Petal Width'),
    split = c(2, 1, 2, 2), more = TRUE)
print(stripplot(Sepal.Length ~ Species, data = iris,
    jitter = TRUE, ylab = '', main = 'Sepal Length'),
    split = c(1, 2, 2, 2), more = TRUE)
print(stripplot(Sepal.Width ~ Species, data = iris,
    jitter = TRUE, ylab = '', main = 'Sepal Width'), split = c(2, 2, 2, 2))
```

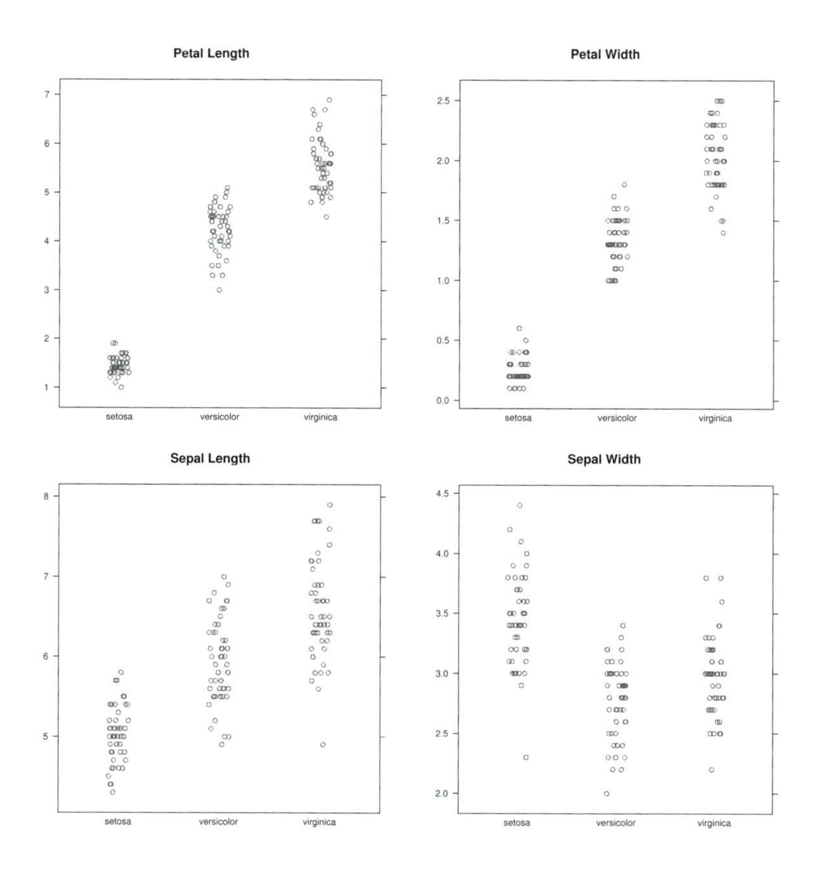

The one-dimensional marginal distributions do not suffice to separate the three groups.

A two-dimensional display does not help much further.

Exercise 4.5	**Iris Classification**
	Use the methods from Section 4.4 and 4.5 to inspect the data set. Can you see classification rules that give a classification of the three species to a large extent?

Using formal methods such as linear discriminant analysis (for example `lda()`, provided in `library(MASS)`), a classification based on the morphometric variables can be found. Separating the species is not trivial.

The original variables represent only an aspect of the data that is technically most accessible. From a biological point of view, a different parametrisation would be chosen. The variables reflect size and form of the leaves. A first approximation with varaiables which are meaningful from a biological point of view would be

$$area = length \cdot width \tag{4.1}$$

$$aspectratio = length/width. \tag{4.2}$$

Variable `area` just catches the size, and `aspectratio` gives a first information about the shape.

Input

```
iris$Sepal.Area <- iris$Sepal.Length*iris$Sepal.Width
iris$Petal.Area <- iris$Petal.Length*iris$Petal.Width
iris$Sepal.Ratio <- iris$Sepal.Length/iris$Sepal.Width
iris$Petal.Ratio <- iris$Petal.Length/iris$Petal.Width

pairs(iris[6:9], main = "Anderson's Iris Data -- 3 species",
      pch = 21,
      bg = c("magenta", "green3", "yellow")[unclass(iris$Species)],
      oma = c(8, 8, 8, 8))
mtext(c("Colour codes:", levels(iris$Species)),
    col = c("black", "magenta", "green3", "yellow"),
    at = c(0.1, 0.4, 0.6, 0.8),
    side = 1, line = 2)
```

In the marginal distribution the species are separated almost completely, with two borderline cases. In these more biological coordinates we see that for classification the area and aspect ratio of the petal are sufficient. Any proposal for a better formal procedure must first improve on this trivial classification rule.

Even an exhaustive search, for example using projection pursuit, only catches projections, that is, it only catches special linear combination of the variables. For the iris data, we first introduced new variables, the areas and aspect ratios. These are non-linear transformations. Only in a second step are the classifying variables identified and the dimension is reduced drastically. In genuinely multivariate problems it is quite common that initially a dimension extension is necessary to solve the problem. Dimension reduction only makes sense if the describing variables set is sufficiently complex to lead to a solution.

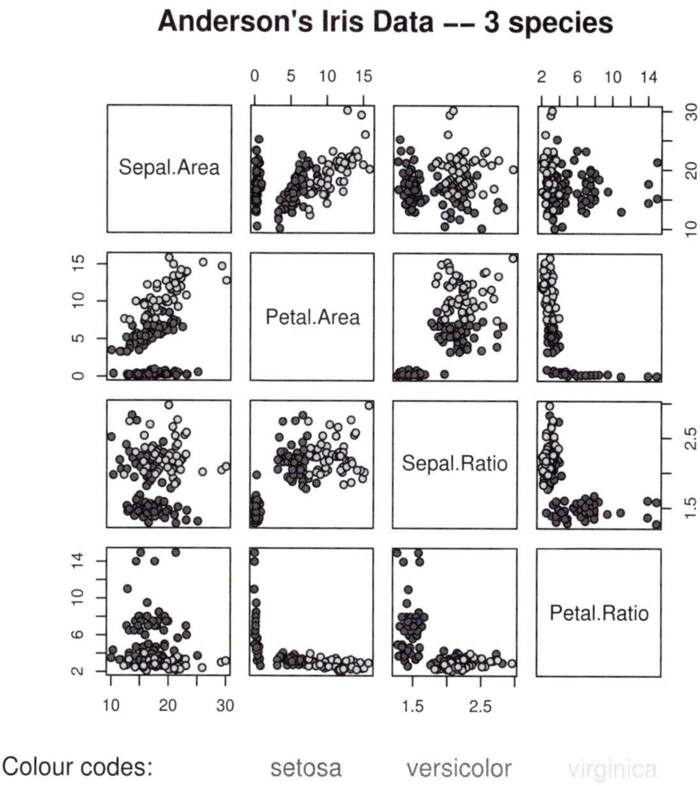

Figure 4.4 *Anderson's Iris data, in transformed coordinates. See Colour Figure 10.*

4.7 Higher Dimensions

4.7.1 Linear Case

If we have essentially linear structures, we can often analyse higher-dimensional structures with methods that have been designed for one-dimensional models. We may have to modify the methods or use them in iterations, taking residuals at each step. But nevertheless they may help us recognise essential features.

In Chapter 2 we introduced linear models for arbitrary regressor dimensions p. However, Chapter 2 presupposes that the model for the statistical analysis is specified off-hand, that is, the information from the data material does not influence the selection of the model, but only decisions within the framework of the model chosen.

In higher-dimensional problems, it is quite common that the model is still to be specified. A common special case is the selection of regressors: the variables are candidates, among which a (preferably small) number of regressors are to be chosen. Do more complicated models provide essential improvement over a simple model? Which parameters resp. which derived variables should be included in the model? The lesson from linear models is that not the values of individual parameters determine the contribution in a model, but that space determined by the parameters is the essential factor. Adapted strategies are needed here. We can start with simple models and ask whether additional parameters make an essential contribution. Additional parameters will improve the fit, but on the other hand the variance of the estimation will increase. We can as well start with a relatively complex model and ask whether we can omit parameters. This may increase the residual error, but we may also gain reliability of the estimations. Both strategies are widely used and are called ***step up strategy*** resp. ***step down strategy***.

For an abstract linear model, both strategies lead to a comparison of two model spaces $\mathscr{M}_{X'} \subset \mathscr{M}_X$. The corresponding estimators are $\pi_{\mathscr{M}_{X'}}(Y)$ and $\pi_{\mathscr{M}_X}(Y)$. The relation between both becomes clear if we choose the orthogonal decomposition $\mathscr{M}_X = \mathscr{M}_{X'} \oplus L_X := M_0, L_X := \mathscr{M}_X \ominus \mathscr{M}_{X'}$ of $\mathscr{M}_X$. Then $\pi_{\mathscr{M}_X}(Y) = \pi_{\mathscr{M}_{X'}}(Y) + \pi_{L_X}(Y)$.

Partial Residuals and Added Variable Plots

In regression $\mathscr{M}_{X'}$ and $\mathscr{M}_X$ are spaces that are spanned by the vectors of regressor variables. In our situation we are interested in the special case

$$X' = \text{span}(X_{1'}, \ldots, X_{p'}); X = \text{span}(X_1, \ldots, X_p)$$

with $p > p'$. Here L_X is spanned by the vectors

$$R_{p'+1} = X_{p'+1} - \pi_{\mathscr{M}'_X}(X_{p'+1}), \ldots, R_p = X_p - \pi_{\mathscr{M}_{X'}}(X_p).$$

If we (formally) do a linear regression of the additional regressors by the regressors already contained in X', the resulting residuals generate L_X. An additional regression of Y by these residuals yields the term $\pi_{L_{X'}}(Y)$, which describes the difference between the models. By construction we know that $\pi_{\mathscr{M}_{X'}}(Y)$ is orthogonal to L_X. So this part is mapped to zero by the second regression. We can as well eliminate this part and restrict ourselves to the regression of $Y' = Y - \pi_{\mathscr{M}_{X'}}(Y)$ after $R_{p'+1}, \ldots, R_p$.

The strategy is simple: we check whether additional parameters should be included in the model. Instead of the scatterplot matrix of the original data we look at the scatterplots of the (formal) residuals from this simple model. These scatterplots are called ***added variable plots***.

To point out the difference from the scatterplot matrix of the original data: linear structures in the scatterplot of the original data are a clear hint for linear dependency. Non-linear structures, such as a triangular shape in some of the scatterplots, can reflect a corresponding dependency structure. But it can also be an artefact, based on the distribution and correlation structure of the regressors. In general, they do not have a simple interpretation. In contrast, the representations in the matrix of the added variable plots are adjusted for linear effects of the preceding variables. As a consequence, they depend on the order in which variables are taken into the model. They correct for linear effects resulting from the correlation of preceding variables. This avoids several artefacts, and the added variable plots are easier to interpret.

Exercise 4.6	
	Modify the following function *pairslm()* so that it calculates the residuals of the regression of all original variables in matrix *x* by regression after the new variable ***x$fit*** and produces a scatterplot matrix of these residuals.
	pairslm <- function(model, x, ...)
	{ x$fit <- lm(model, x)$fit; pairs(x, ...) }
	Add title, legends, etc.
	Use the "trees" data set as an example.

We discussed the transition from p' to $p'+1$ variable. The scatterplot matrix allows us to get a quick overview of a (not too large) number of candidates (three potential additional regressors in our case). The transition from p to $p-1$ for the elimination of a variable is in a certain sense dual to this. It corresponds to stepwise elimination, or step down, the other basic strategy.

Instead of using individual variables step by step, it can be more efficient to select linear combinations of several variables and use these synthetic variables in a model. Methods to achieve this are discussed under the heading ***principal component analysis*** and supported by function *prcomp()*. We will return to this in a later example (page 185).

The experience with the linear models teaches us that the marginal relations are only half of the truth. Instead of using the regressors individually, we have to use a stepwise orthogonalisation. Component-wise interpretation of estimators becomes questionable. The estimation may depend on the order in which regressors are used.

In more complex situations, formal methods may lead astray. Technical skills are needed. Unfortunately, knowledge about how interventions affect the validity of formal methods is rather limited. This makes it more important to judge chosen strategies critically using simulations.

4.7.2 Non-Linear Case

Non-linear relations in higher dimensions pose a challenge. Besides a good collection of methods, we need a repertoire of examples showing which structures can occur and where we have to pay attention. The following example, the cusp singularity, is one of these examples. It is one of the simplest structures that can occur in higher dimensions. The base here is a two-dimensional structure, a surface, that is not trivially imbedded in a three-dimensional space. The interesting feature here is the bifurcation from a unimodal to a bimodal situation.

Example: Cusp Non-Linearity

The simplest example can be illustrated with reference to physical applications. In physical systems, probability distributions often are related to energy levels; (local) minima of the energy correspond to modes of the distribution. A typical relation is if the energy behaves as $\varphi(y)$, after standardisation the distribution behaves like $e^{-\varphi(y)}$. If $\varphi(y)$ is quadratic at the minimum, up to scale transformation, we get distributions from the normal distributions family.

Differential topology teaches us that the qualitative image persists under small perturbations or variations. At least locally the energy stays approximatively quadratic, and the normal distributions stay at least approximatively a suitable distribution family.

The behaviour changes drastically if the behaviour behaves locally like y^4. Small perturbations can make the potential locally a quadratic function. But they can as well lead to breaking the local minimum into two minima. The typical image has the form

$$\varphi(y; u, v) = y^4 + u \cdot y^2 + v \cdot y. \tag{4.3}$$

The variations are controlled by the parameters u, v. A dynamic interpretation helps us

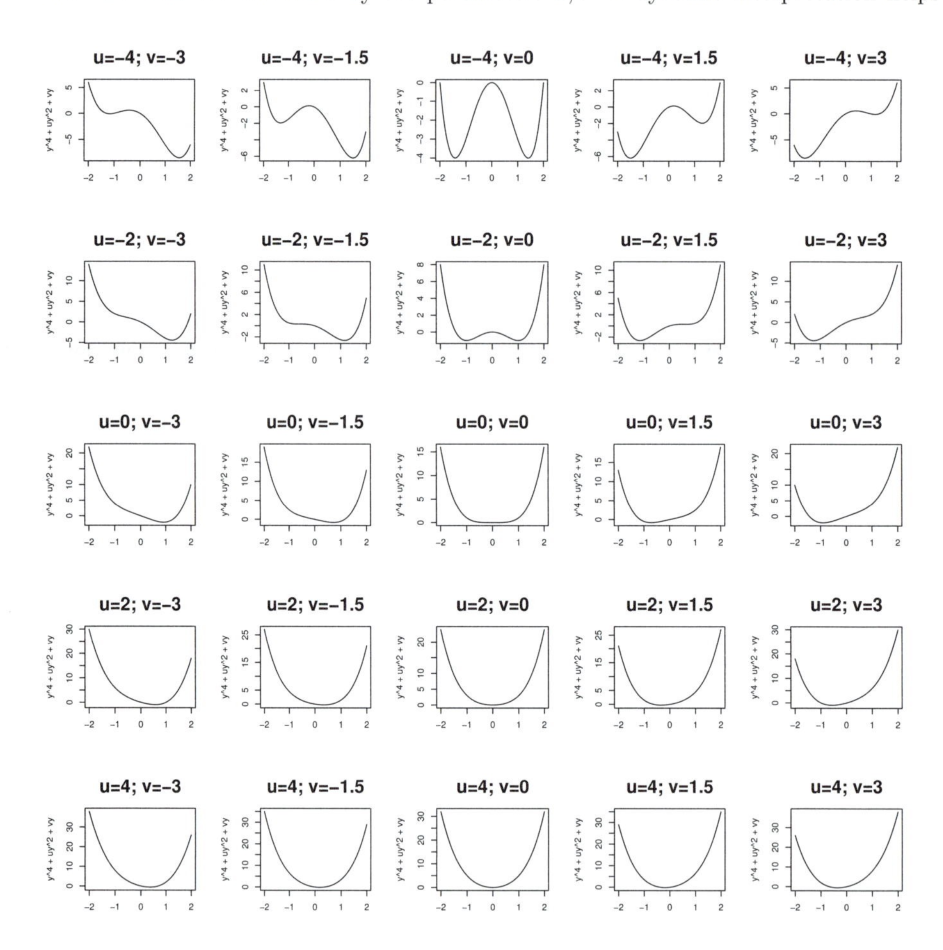

Figure 4.5 *Cusp bifurcation. Unfolding of* y^4: $\varphi(y; u, v) = y^4 + u \cdot y^2 + v \cdot y$

understand the situation: imagine that u, v are external parameters that can change. We know this scenario from magnetic hysteresis: y gives the magnetisation in a certain direction, u takes the role of temperature; v that of an external magnetic field. At high temperature, magnetisation follows the external magnetic field directly. At lower temperatures, the material shows memory. The magnetisation does not only depend on the outer magnetic field, but also on the previous magnetisation.

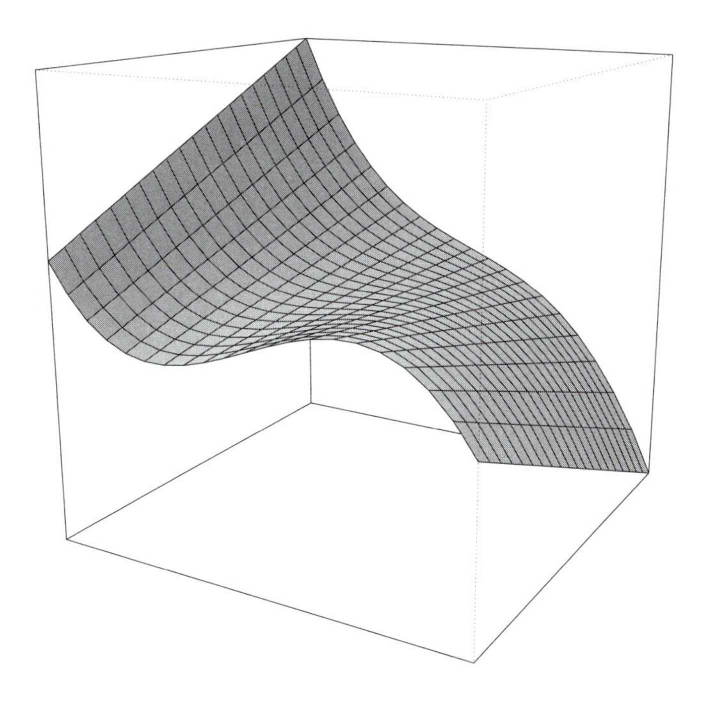

Figure 4.6 *Cusp bifurcation, Critical points* $\varphi'(y; u, v) = 4y^3 + 2u \cdot y + v = 0$

We know similar "memory effect" in other areas. Imagine a market with prices y, costs v and a "pressure of competition" u. With sufficient competition the prices will follow the cost, more or less, given other conditions are unchanged. In a monopoly situation, prices seem to have a memory: once they have risen, they will only become lower if the costs reduce drastically.

The "unfolding" of the potential y^4 given in Formula 4.3 has a characteristic form. From

$$\varphi'(y; u, v) = 4y^3 + 2u \cdot y + v = 0 \tag{4.4}$$

we get the critical points (see Figure 4.6).

A projection to the u, v plane gives a cusp (Figure 4.7). For parameter values in the interior of this cusp the potential has two local minima. Outside of the cusp there is only one extreme value.

The distributions corresponding to these potentials are, up to scale transformation for normalisation,

$$p(y; u, v) \propto e^{-(y^4 + u \cdot y^2 + v \cdot y)}. \tag{4.5}$$

The structure of the potentials carries over to the corresponding distributions.

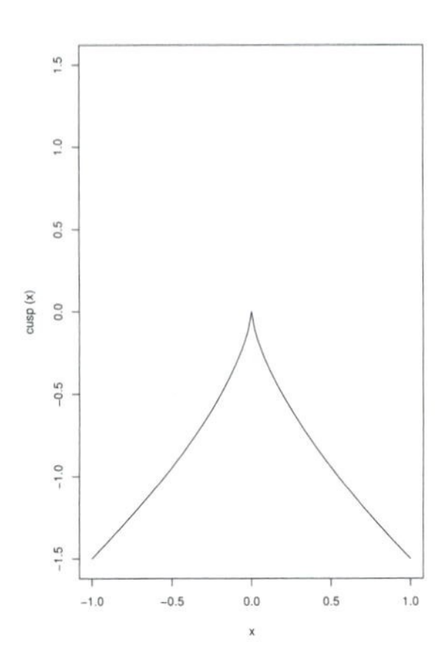

Figure 4.7 *Cusp bifurcation. Boundary between uni- and bimodality in (u, v) space*

The situation appears mostly harmless: the parameter space (the space of the regressors) $x = (u, v)$ has only two dimensions. The distribution is one-dimensional with a smooth density. But the situation cannot be captured sufficiently by linear methods. You will not recognise the typical non-linear effect if you are not prepared for it. Only an overall picture in three dimensions will convey the underlying structure.

This simple example is a challenge. How can we diagnose a structure of this kind?

Exercise 4.7	**Cusp Bifurcation**
	Write a function `dx4exp(x, u, v)` to calculate the centred probability densities for (4.5). You have to integrate the density from (4.5) to determine the norming constant and to calculate the expected value to centre the density. For both, you can use a numerical integration with `integrate()`.
***	For values u, v on a grid in $u = -2 \ldots 2$ and $v = -1 \ldots 1$ simulate 100 random numbers from `dx4exp(x, u, v)`. Use the methods from Chapter 2 to analyse the data. Can you detect hints indicating the non–linear structure? Is the bi-modality recognisable? How much of the structure can you identify?

In non-linear relations, joint dependency can be very important. In general, this requires prudence in modelling. Non-linear relations may be hidden in projections. Artefacts of the (linear) projection can give an image that does not correspond to the original relations.

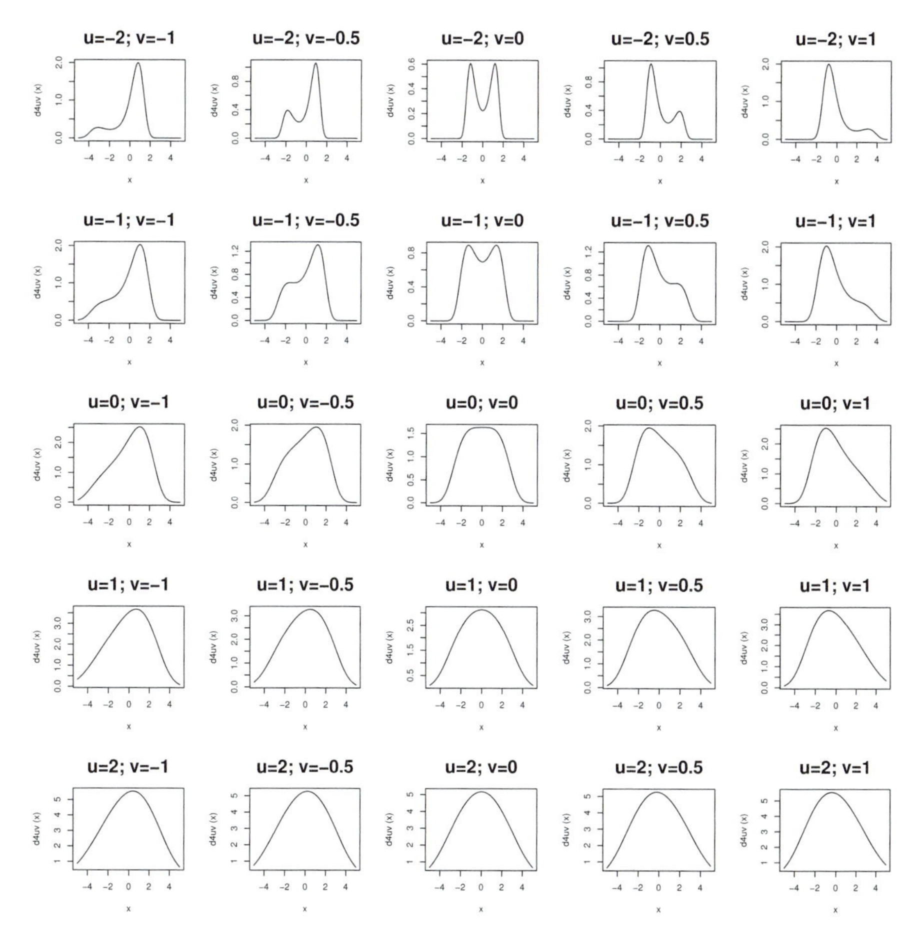

Figure 4.8 *Cusp bifurcation:* $p(y; u, v) \propto e^{-(y^4+u \cdot y^2+v \cdot y)}$

4.7.3 Case Study: Melbourne Temperature Data

R. Hyndman pointed out the bifurcation to bimodality in the Melbourne temperature data set [16]. We use an extended version of the data set[2] and analyse the day-by-day difference in temperature at 15h (the daily report reference time) conditioned on today's temperature and pressure at the reference time.

We are interested in modality, and classical methods like those introduced in Chapter 1 are not tuned for uncovering multi-modality. One reason is that modality is a multi-scale problem. If we use for example kernel density estimation (see Section 1.4), typically we get exactly one mode for sufficiently large bandwidth, and a mode at every data point if the bandwidth is sufficiently low. To understand modality we have to look at different scales and find the suitable ones.

Instead of going into this problem of bandwidth choice, we use the shorth plot, which allows

[2] Melbourne temperature data 1955–2007, provided by the Bureau of Meteorology, Victorian Climate Services Centre, Melbourne.

a multi-scale representation. For each data point x, the plot shows the minimum length of an interval containing x and covering at least a proportion α of the data. Using this plot, we can investigate several levels of α simultaneously. For details and source code, see `<http://lshorth.rforge.r-project.org/>` [45].

The shorth plot view is in Figure 4.9, showing the split into bimodality. The full picture reveals a cusp-type bifurcation. Figure 4.10 shows the shorth plot for the temperature difference at 15h (the daily report reference time) to the next day's temperature, conditioned on today's temperature and pressure. It reveals the modality split as well as the skewness, which is pressure dependent.

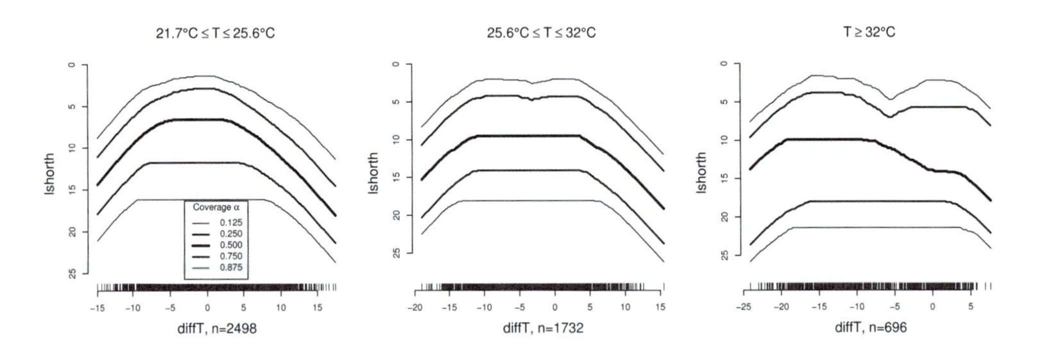

Figure 4.9 *Shorth plot at coverage levels* $\alpha = 0.125, 0.25, 0.5, 0.75, 0.875$ *for Melbourne day-by-day temperature difference at 15:00h conditioned at today's temperature. A bifurcation to bimodality occurs at high temperatures.*

4.7.4 Curse of Dimensionality

With the lack of adapted coordinate systems, an extensive search for interesting projections or sections is necessary. The number of possibilities increases rapidly with the dimension. For illustration: to identify a cube, at least the corner points must be detected. In d dimensions there are 2^d corner points. The number increases exponentially with the dimension. This is an aspect of the problem known as the ***curse of dimensionality***.

Seen from another point of view: if we look at the data points that are extreme in at least one variable dimension, in the one-dimensional case we get two points. In d dimensions typically we get 2^d extreme points. If we do not look at coordinate directions, but in arbitrary directions, typically any point is extreme if d gets large.

A third aspect: in d-dimensional space almost every point is isolated. Localisations, like those discussed in Section 2.5, will break down. If we take a neighbourhood around some point covering a proportion p of the data range, for example $p = 10\%$ of the span of the variables, in one dimension typically we cover a proportion of p of the data points. In d dimensions this is only a fraction in the order of p^d. So for example in 6 dimensions we need several million data points in order to have non-empty environments for most data points.

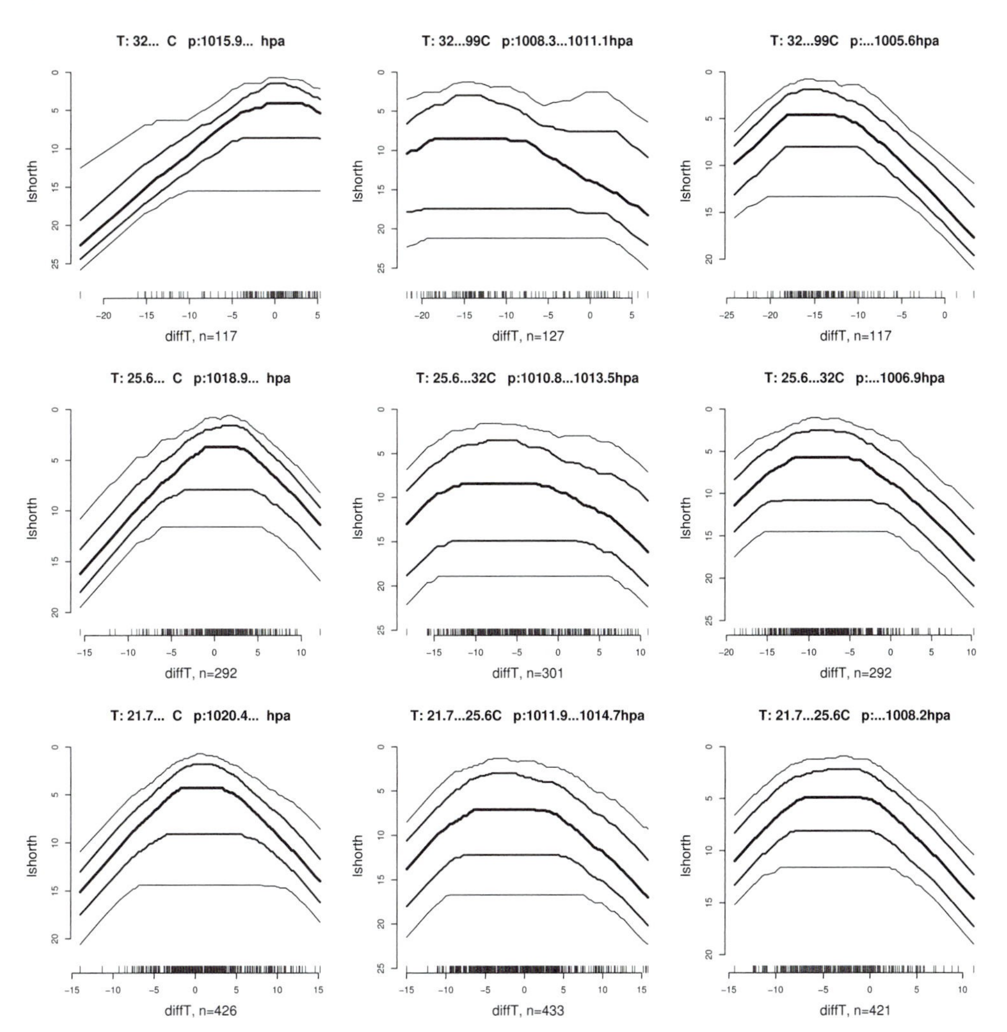

Figure 4.10 *Plot matrix of shorth plots for varying temperature ranges (vertical) and varying pressures (horizontal). Shorth plots at coverage levels* $\alpha = 0.125, 0.25, 0.5, 0.75, 0.875$ *for Melbourne day-by-day temperature difference at 15:00h, conditioned at today's temperature and pressure.*

4.7.5 Case Study: Body Fat

As a running example we use here the fat data set [21]. This data set is published repeatedly in the literature and in R among others available in the library *UsingR* [53].

The target of the investigation related to this data set is the determination of the proportion of body fat. The most reliable method is to use a water bed to determine the average density of the tissue and to infer from the density to the body fat proportion. This measurement is

Name	Variable	Unit, Remarks
case	case number	
body.fat	percent body fat using Brozek's equation, $457/density - 414.2$	
body.fat.siri	percent body fat using Siri's equation, $495/density - 450$	
density	density	$[g/cm^2]$
age	age	$[yrs]$
weight	weight	$[lbs]$
height	height	$[inches]$
BMI	adiposity index = $weight/height^2$	$[kg/m^2]$
ffweight	fat-free weight = $(1-fraction\ of\ body\ fat)*$ $weight$, using Brozek's formula	$[lbs]$
neck	Neck circumference	$[cm]$
chest	Chest circumference	$[cm]$
abdomen	Abdomen circumference "at the umbilicus and level with the iliac crest"	$[cm]$
hip	hip circumference	$[cm]$
thigh	thigh circumference	$[cm]$
knee	knee circumference	$[cm]$
ankle	ankle circumference	$[cm]$
bicep	extended biceps circumference	$[cm]$
forearm	forearm circumference	$[cm]$
wrist	wrist circumference "distal to the styloid processes"	$[cm]$

Table 4.15 *Fat data set: variables*

rather intricate. Can it be replaced by body parameters that allow a simpler measurement? The parameters at disposition are summarised in Table 4.15.

The survey in Table 4.15 tells us that metric variables and US scales are mixed. To make interpretation easier for us, we convert all information to metric units.

Input

```
library("UsingR")
data(fat)
fat$weightkg <- fat$weight*0.453
```

```
fat$heightcm <- fat$height * 2.54
fat$ffweightkg <- fat$ffweight*0.453
```

The variables *body.fat* and *body.fat.siri* are derived from the observed values of *density*. These formulas reflect assumptions about the mean density of fat and of fat-free tissue. With these assumptions the fat proportion can be calculated (or rather: estimated) from the *density*. Both formulas use a density dependent factor *1/density*. Up to (given or assumed) constants this (and not the *density*) is the relevant term for us.

The first step is a critical inspection and cleaning of the data set. This is almost always necessary, not only for higher-dimensional data sets. In higher-dimensional data sets, though, we often have redundancies that allow for consistency checks and possibly for corrections. In our case, the variables *body.fat*, *body.fat.siri*, *ffweight* and *BMI* are derived variables that are in a deterministic relation to other variables.

We first look at the group *body.fat, body.fat.siri, 1/density*. The pair-wise scatterplots should show a straight line. *pairs()* provides good service. We use the formula-based notation here. To mark that *1/density* should be calculated, and the / operator is not to be understood as a formula operator, we have to mark up this term.

Input

```
pairs(~body.fat + body.fat.siri + I(1/density), data = fat)
```

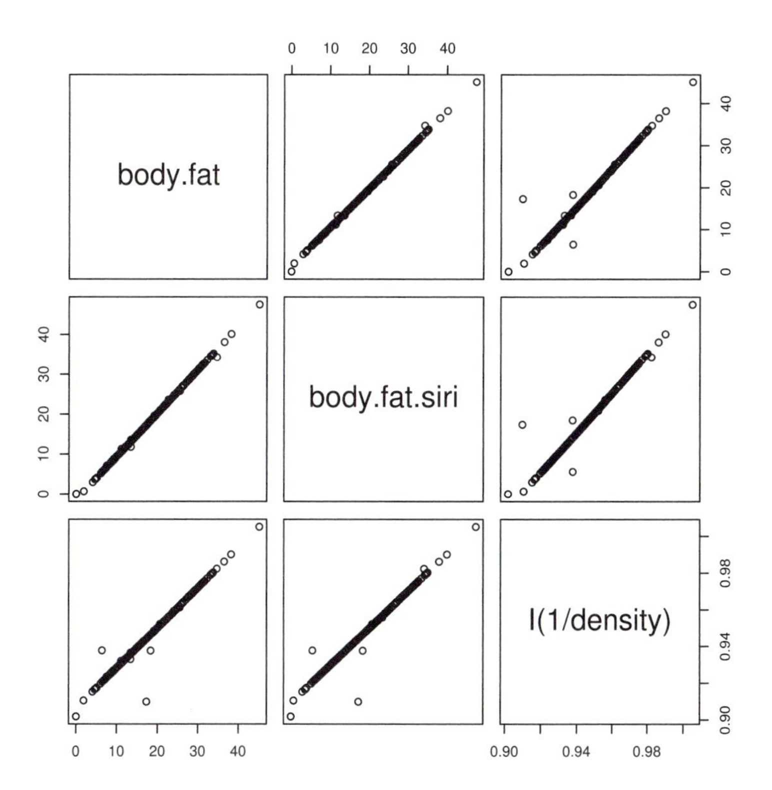

The inconsistent values and outliers are obvious. Unfortunately, in R it is not simple to mark values in a scatterplot matrix.[3]

Exercise 4.8

Use functions *plot(). identify() text.id()*, to generate the following output:

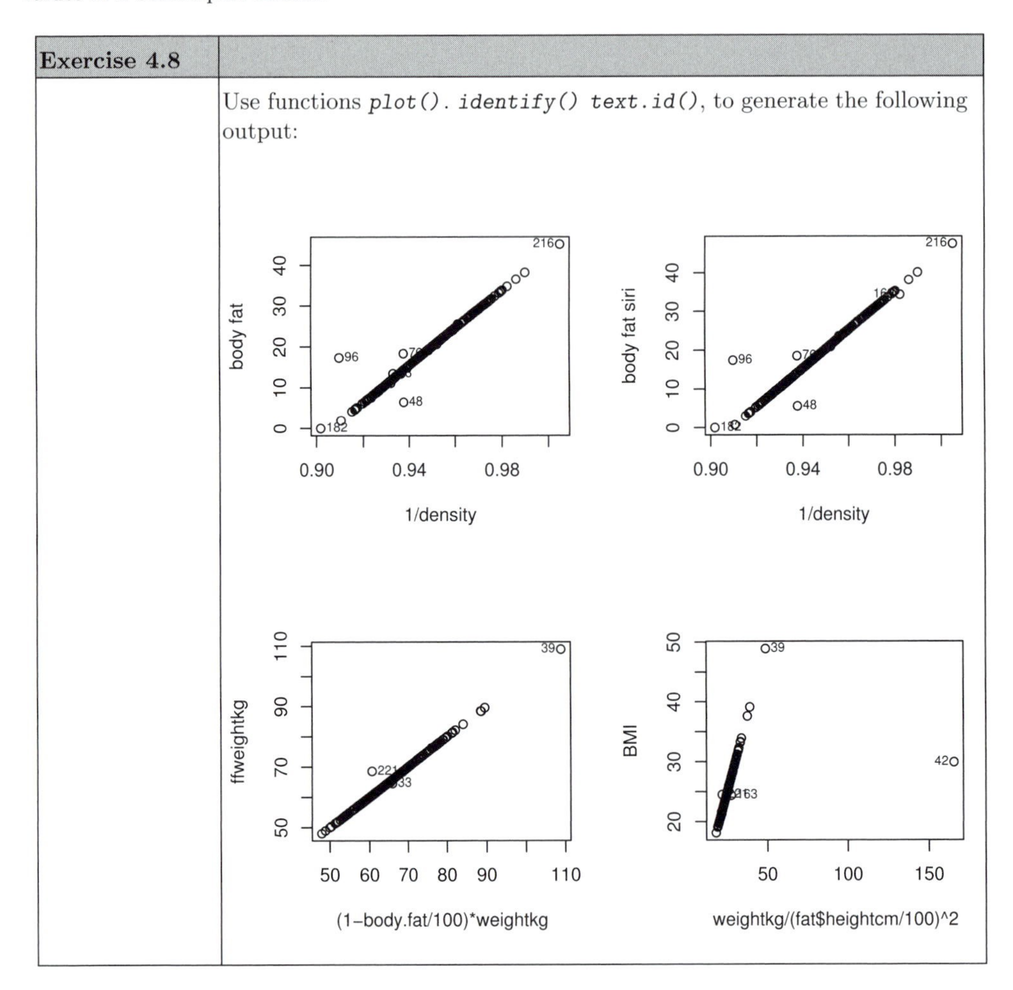

If an obvious correction is possible, it should be done here. Of course the correction needs to be noted in the report of the analysis. Case 42 is simple: a height of 0.73m for a weight 63.5kg is not plausible and not consistent with BMI 29.9. From the BMI we can recalculate the height. Instead of the value of 29.5 inches the height should presumably be 69.5 inches.

Input

```
fat$height [42] <- 69.5
fat$heightcm[42] <- fat$height[42] * 2.54
```

Case 216 is discretionary. The density is extremely low, the BMI extremely high. On the other

[3] This applies to the basic R version. Add-on packages or external programs which allow more flexibility, such as ggobi, are available.

hand, the body measurements fit to these extremes. This case can be an outlier, which would distort the analysis. But it could also be an observation that is particularly informative. We note it as a particularity.

After this preliminary inspection we clean up the data set. We remove the variables that contain no additional information or have been replaced. As a target variable, we use `body.fat`. However, we keep the variable `density` for later purposes.

Input

```
fat$weight <- NULL
fat$height <- NULL
fat$ffweight <- NULL
fat$ffweightkg <- NULL
fat$body.fat.siri <- NULL
```

Some indices are commonly used to describe the body constitution. Some time ago, the rule of thumb was $optimalweight[kg] = height[cm] - 100$. Today the "body mass index" $BMI = weight/height^2$ is usual (see Figure 4.11). Today, commercially available scales determine the body fat by electrical impedance. This variable is not contained in the fat data.

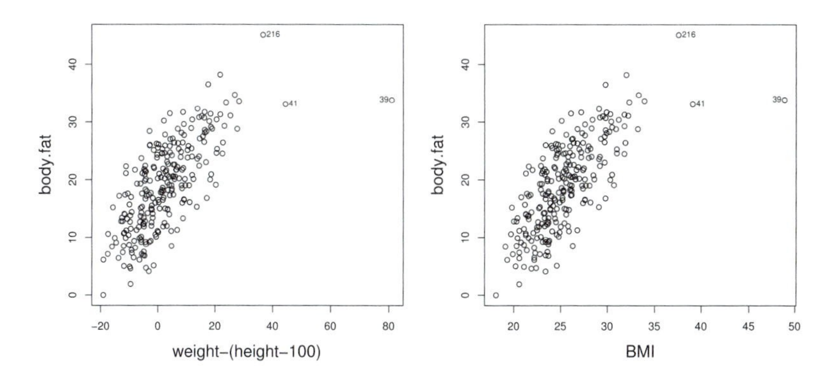

Figure 4.11 *Body fat against conventional indices*

We can use the conventional indices in a linear model. We exclude the obvious outliers and possible leverage points. This can be achieved with the `subset` parameter of `lm()`.

For the rule of thumb $optimalweight = height - 100$ we get:

Input

```
lm.height <- lm(body.fat~I(weightkg-(heightcm-100)),
    data = fat,
    subset = -c(39, 41, 216))
summary(lm.height)
```

Output

```
Call:
lm(formula = body.fat ~ I(weightkg - (heightcm - 100)), data = fat,
    subset = -c(39, 41, 216))

Residuals:
     Min        1Q    Median        3Q       Max
-11.90734  -3.68697  -0.05303   3.65458  12.28000

Coefficients:
                               Estimate Std. Error t value Pr(>|t|)
(Intercept)                    17.70722    0.33296   53.18   <2e-16 ***
I(weightkg - (heightcm - 100))  0.54557    0.03283   16.62   <2e-16 ***
---
Signif. codes:  0 '***' 0.001 '**' 0.01 '*' 0.05 '.' 0.1 ' ' 1

Residual standard error: 5.166 on 247 degrees of freedom
Multiple R-squared: 0.5279,        Adjusted R-squared: 0.526
F-statistic: 276.2 on 1 and 247 DF,  p-value: < 2.2e-16
```

The regression of *body.fat* for *BMI* results in:

Input

```
lm.BMI <- lm(body.fat~BMI,
    data = fat,
    subset = -c(39, 41, 216))
summary(lm.BMI)
```

Output

```
Call:
lm(formula = body.fat ~ BMI, data = fat, subset = -c(39, 41,
    216))

Residuals:
     Min        1Q    Median        3Q       Max
-12.49460  -3.53561  -0.05228   3.69129  11.72720

Coefficients:
            Estimate Std. Error t value Pr(>|t|)
(Intercept) -25.6130     2.6212  -9.772   <2e-16 ***
BMI           1.7564     0.1031  17.042   <2e-16 ***
---
Signif. codes:  0 '***' 0.001 '**' 0.01 '*' 0.05 '.' 0.1 ' ' 1

Residual standard error: 5.097 on 247 degrees of freedom
Multiple R-squared: 0.5404,        Adjusted R-squared: 0.5385
F-statistic: 290.4 on 1 and 247 DF,  p-value: < 2.2e-16
```

The fit of $R^2 = 0.53$ resp. $R^2 = 0.54$ is poor in both cases.

Even using all data points and all regressors, we get at most $R^2 = 0.75$:

Input

```
lm.fullres <- lm(body.fat ~ age + BMI + neck + chest +
    abdomen + hip + thigh + knee + ankle +
    bicep + forearm + wrist + weightkg + heightcm,
    data = fat)
summary(lm.fullres)
```

Output

```
Call:
lm(formula = body.fat ~ age + BMI + neck + chest + abdomen +
    hip + thigh + knee + ankle + bicep + forearm + wrist + weightkg +
    heightcm, data = fat)

Residuals:
     Min       1Q   Median       3Q      Max
-10.0761  -2.6118  -0.1055   2.8993   9.2691

Coefficients:
              Estimate Std. Error t value Pr(>|t|)
(Intercept) -50.804727  36.489198  -1.392  0.16513
age           0.061005   0.029862   2.043  0.04217 *
BMI           0.782993   0.733562   1.067  0.28688
neck         -0.439082   0.218157  -2.013  0.04528 *
chest        -0.040915   0.098266  -0.416  0.67751
abdomen       0.866361   0.085550  10.127  < 2e-16 ***
hip          -0.206231   0.136298  -1.513  0.13159
thigh         0.246127   0.135373   1.818  0.07031 .
knee         -0.005706   0.229564  -0.025  0.98019
ankle         0.135779   0.208314   0.652  0.51516
bicep         0.149100   0.159807   0.933  0.35177
forearm       0.409032   0.186022   2.199  0.02886 *
wrist        -1.514111   0.493759  -3.066  0.00242 **
weightkg     -0.389753   0.221592  -1.759  0.07989 .
heightcm      0.187196   0.199854   0.937  0.34989
---
Signif. codes:  0 '***' 0.001 '**' 0.01 '*' 0.05 '.' 0.1 ' ' 1

Residual standard error: 3.991 on 237 degrees of freedom
Multiple R-squared: 0.7497,      Adjusted R-squared: 0.7349
F-statistic:  50.7 on 14 and 237 DF,  p-value: < 2.2e-16
```

This is a model with 15 coefficients. The model is so complex that it is difficult to interpret, and one should try to reduce the model. Instead of searching "by hand" for simpler models, this process can be automated. This is done for example by function `regsubsets()` in `library(leaps)` [49]. The quadratic error (resp. the coefficient of determination R^2) has to be modified: the quadratic error is trivially minimised if we take all regressors into the model, so the complete model will always be the optimal model in terms of the coefficient of determination R^2. For model selection one uses variants of the quadratic error (resp. the coefficient of determination R^2) that are adjusted for the number of parameters.

Exercise 4.9	

*	Use
	`library(leaps)`
	`lm.reg <- regsubsets(body.fat ~ age + BMI + neck + chest + abdomen + hip + thigh + knee + ankle + bicep + forearm + wrist + weightkg + heightcm, data = fat)}`
	and inspect the result with
	`summary(lm.reg)`
	`plot(lm.reg, scale = "r2")`
	`plot(lm.reg, scale = "bic"`
	`plot(lm.reg, scale = "Cp")`
	Hint: See `help(plot.regsubsets)`.

*	Use the function `leaps()` for model selection.

But the tools that we discussed in Chapter 2 are tarnished now. The statistical statements in the summary are only valid, if model resp. hypothesis are stated independent of the data material. If a model is data based, the distribution of the estimated coefficients is unclear. We do not know how to define confidence intervals or how to perform tests. The software does not know that we are in a process of model selection and just returns the probabilities that apply for a fixed pre-selected model under the assumption of a normal distribution.

The diagnostics, such as distribution plots of the residuals, becomes useless as well: the central limit theorem ensures that under weak independence conditions the residuals approximately have a normal distribution if we have a large number of terms, even if the normal distribution assumption does not hold for the errors themselves.

We are at a dead end.

We illustrate a different approach that goes further. To do this we return to the beginning of the analysis, after the preliminary inspection and data correction. To avoid running into the problem that the statistical distributions are influenced by preceding model selection steps, we split the data set. We use one part as a training set for model selection and to play with various alternatives. The rest is reserved as an evaluation set. The information from this part is only used after model selection for the statistical analysis.

A closer look shows that the model selection is most critical for the estimation of the error, not so much for parameter estimation. If the error is estimated based on the data used for model selection, we tend to underestimate the error. The evaluation part serves as a reliable estimation of the error and for residual diagnostics. This is a limited task. We reserve only a smaller proportion of the data for this.

Input

```
sel <- runif(dim(fat)[1])
fat$train <- sel < 2/3
rm(sel)
```

We discard the outliers from the training part in case they are included in our random selection.

Input

```
fat$train[c(39, 41, 216)] <- FALSE
summary(fat$train)
```

Output

```
   Mode   FALSE    TRUE    NA's
logical      82     170       0
```

Our target variable is `body.fat`, or alternatively `1/density`.

We try to structure the variables with respect to the subject matter. For the density we have a definition from physics

$$density = \frac{weight}{volume}.$$

Among the variables that are regressor candidates there is just one variable that directly refers to weight (`weight` resp. `weightkg`), one variable for `age`, and a whole series of variables referring to body geometry.

From the observed density and the observed weight we can retrieve the volume. We add this to our variable set. Since we have just one weight observation per person, there is no space for person-related statistics.

Input

```
fat$vol <- fat$weightkg/fat$density
```

Next we try to model the volume `fat$vol` in terms of the more convenient variables on the body geometry. The values reported to describe the body geometry represent linear measurements. In a rough approximation, we can derive volume related variables. The only length information we have is in `height`. For the lack of better information we assume that all parts of the body have a length that is proportional to the height. We are aiming for a linear model. Hence linear factors can be neglected since they are estimated by the model anyway. We use a rough model for the body, similar to a member doll that artists use. If we take cylindrical approximations for the parts of the body, up to linear factors we have:

Input

```
fat$neckvol <- fat$neck^2 * fat$heightcm
fat$chestvol <- fat$chest^2 * fat$heightcm
fat$abdomenvol <- fat$abdomen^2 * fat$heightcm
fat$hipvol <- fat$hip^2 * fat$heightcm
fat$thighvol <- fat$thigh^2 * fat$heightcm
fat$kneevol <- fat$knee^2 * fat$heightcm
fat$anklevol <- fat$ankle^2 * fat$heightcm
fat$bicepvol <- fat$bicep^2 * fat$heightcm
fat$forearmvol <- fat$forearm^2 * fat$heightcm
fat$wristvol <- fat$wrist^2 * fat$heightcm
```

We need not worry about the constants – they are estimated by the linear model anyhow.

Next we inspect the internal structure of the volume-oriented regressor candidates, using the training data set only. “Internal structure” here means that we are just investigating the relation of the regressor candidates among one another, leaving the response out of discussion for now. For a first graphical representation, we use parallel coordinates. Since the display is strongly influenced by the sequence and the layout of the display space, we use a standard trick and show the variables of interest twice. We make use of the alpha channel again to resolve overplotting.

***Example* 4.4: Parallel Plots**

Input

```
print(parallel(fat[ -c(39, 41, 216), c(19:29,19:29)],
    horizontal.axis = FALSE, scales = list(x = list(rot = 90)),
    col =  rgb(red=0, blue=0, green=0, alpha=0.2))
)
```

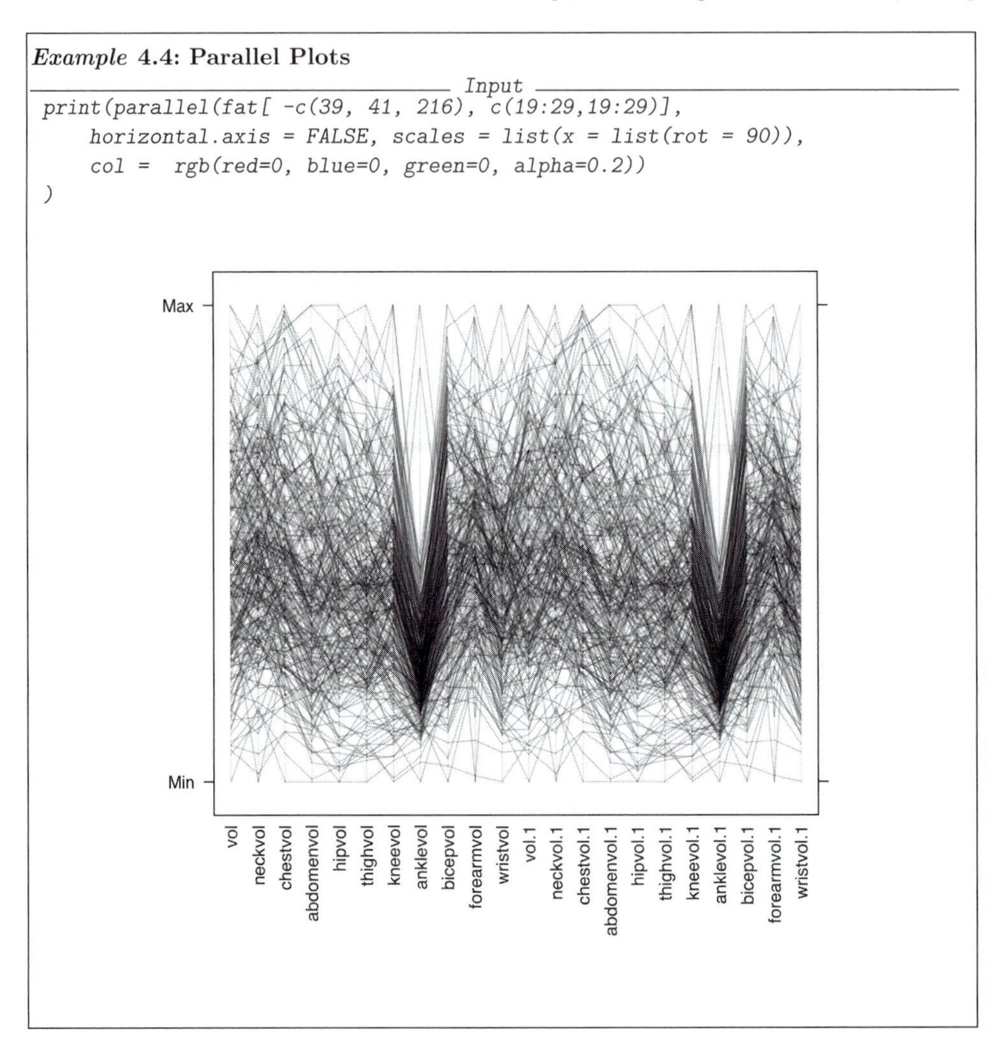

Parallel coordinates require some training. But what is obvious at first sight is that there is a particular internal structure linking the volumes attributed to biceps and knee on the one hand, and ankle on the other hand. And, in case we did not note it before, the parallel coordinates plot reveals that there are two cases with exceptionally high values for ankle.

Exercise 4.10	
	Remove the obvious outliers and rearrange the variables starting from the body volume so that on average the correlation between subsequent variables is maximized.

We use the function *prcomp()*, which for any given set of variables yields stepwise best linear predictors.

For the approximative volumes of the body parts, the principal components are:

Input
```
pcfatvol <- prcomp(fat[, 20:29], subset = fat$train)
round(pcfatvol$rotation, 2)
```

Output
```
            PC1   PC2   PC3   PC4   PC5   PC6   PC7   PC8   PC9  PC10
neckvol    0.05 -0.02  0.06 -0.07  0.57  0.05  0.79 -0.17  0.07  0.09
chestvol   0.55  0.46  0.69  0.01 -0.11  0.00  0.00 -0.01  0.01  0.00
abdomenvol 0.65  0.29 -0.70 -0.06  0.05 -0.01 -0.02  0.02 -0.02  0.00
hipvol     0.48 -0.75  0.10  0.44 -0.03  0.07  0.00  0.02  0.01  0.00
thighvol   0.18 -0.37  0.07 -0.87 -0.24 -0.02  0.08  0.01 -0.02 -0.02
kneevol    0.06 -0.08  0.06 -0.06  0.34 -0.82 -0.28 -0.16  0.30  0.06
anklevol   0.02 -0.03  0.03 -0.01  0.14 -0.24 -0.05 -0.12 -0.95  0.07
bicepvol   0.05 -0.05  0.08 -0.18  0.55  0.51 -0.53 -0.33  0.03  0.01
forearmvol 0.02 -0.02  0.07 -0.09  0.39  0.02 -0.10  0.91 -0.05  0.04
wristvol   0.01 -0.01  0.02  0.00  0.11 -0.05  0.04  0.00 -0.05 -0.99
```

The pattern of the signs for the loadings gives hints about the internal structure. The first principal component *PC1* is a linear combination of variables which essentially describes the torso. The second principal component contrasts the upper part of the torso (chest to abdomen) with the lower torso. The third contrasts the abdomen with the rest of the torso.

So in our context, the first three components are easy to interpret. To understand the other components, we can get a graphical representation of the principal components by showing the variable axes as they appear projected to the space spanned by the principal components. This representation is called a ***biplot***. We use it for the next two principal components $PC4$ and $PC5$. Since the plotting area is a limited space, we do use abbreviated variable names. This kind of direct intervention is error prone if done ad hoc. We use a software solution for this step.

Input
```
abbreviate(rownames(pcfatvol$rotation),4)
```

Output
```
   neckvol   chestvol abdomenvol     hipvol   thighvol    kneevol
    "nckv"     "chst"     "abdm"     "hpvl"     "thgh"     "knvl"
  anklevol   bicepvol forearmvol   wristvol
    "ankl"     "bcpv"     "frrm"     "wrst"
```

***Example* 4.5: Biplot**

Input

```
biplot(pcfatvol, choices = 4:5,
    col = c("grey50", "red"),
    ylabs = abbreviate( rownames(pcfatvol$rotation), 4 ),
    cex = c(0.8, 1.2))
```

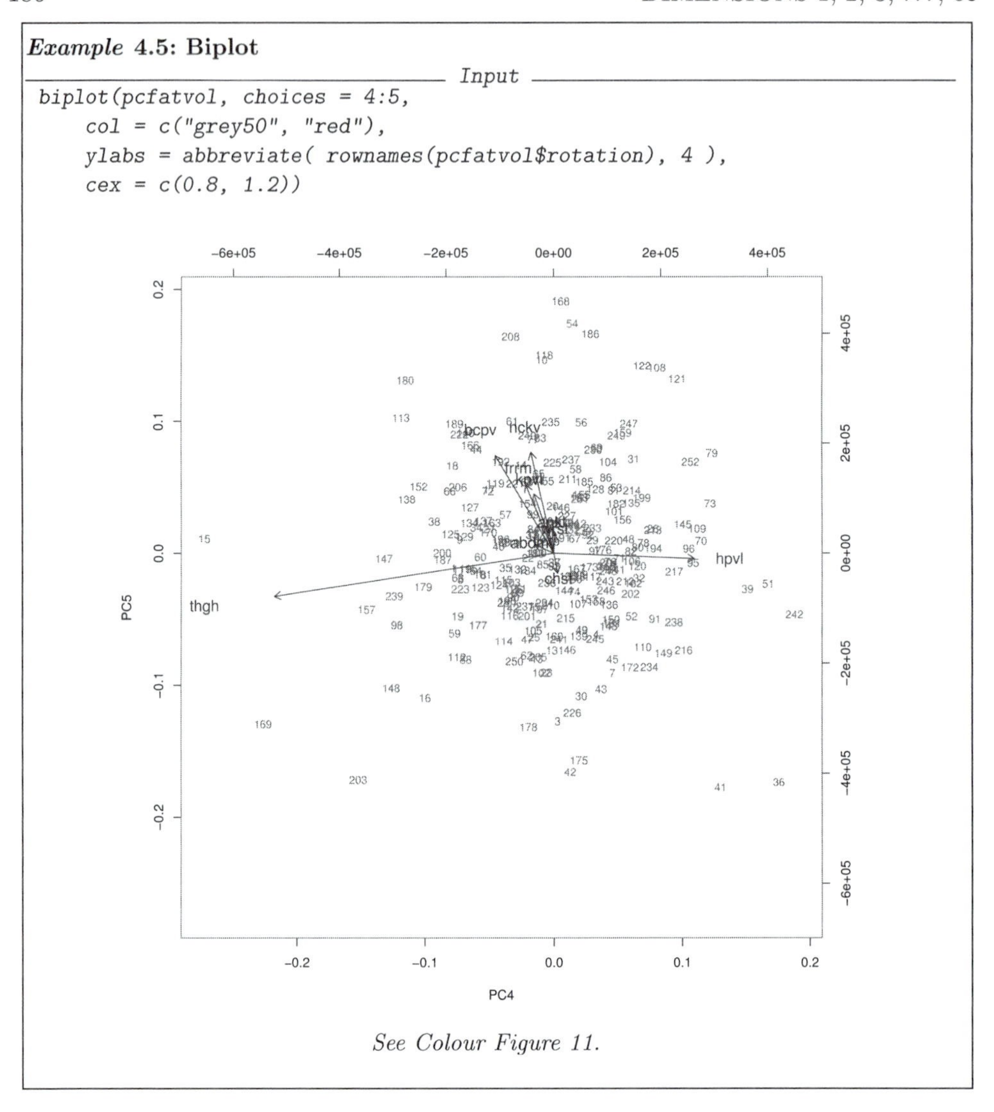

See Colour Figure 11.

However, as is often the case, some thought about the subject matter may help more than formal analysis. We saw this point in the iris example in Section 4.6 (page 162) already.

Exercise 4.11	**Think !**
	Draw a sketch of a member doll that shows which body geometry features are represented by the next principal component *PC4*, ..., *PC10*. For a start, you can concentrate on the signs of the variable weights.

The attempt to represent the derived total volume by the approximative volumes of the body parts results in a high coefficient of determination R^2.

Input
```
lm.vol <- lm(vol ~ neckvol + chestvol + abdomenvol +
    hipvol + thighvol + kneevol +
    anklevol + bicepvol + forearmvol +
    wristvol,
    data = fat, subset = fat$train)
summary(lm.vol)
```

Output
```
Call:
lm(formula = vol ~ neckvol + chestvol + abdomenvol + hipvol +
    thighvol + kneevol + anklevol + bicepvol + forearmvol + wristvol,
    data = fat, subset = fat$train)

Residuals:
    Min      1Q  Median      3Q     Max
-9.4942 -1.0291  0.1805  1.0389  4.5095

Coefficients:
              Estimate Std. Error t value Pr(>|t|)
(Intercept)  3.400e+00  1.389e+00   2.448 0.015469 *
neckvol      1.201e-05  8.638e-06   1.391 0.166235
chestvol     8.401e-06  1.316e-06   6.385  1.8e-09 ***
abdomenvol   1.302e-05  1.117e-06  11.653  < 2e-16 ***
hipvol       7.072e-06  1.834e-06   3.855 0.000167 ***
thighvol     1.278e-05  3.218e-06   3.972 0.000108 ***
kneevol      9.261e-06  9.651e-06   0.960 0.338672
anklevol     2.604e-05  1.305e-05   1.996 0.047648 *
bicepvol     3.069e-05  9.104e-06   3.371 0.000939 ***
forearmvol   3.571e-05  1.283e-05   2.783 0.006035 **
wristvol    -3.424e-05  4.122e-05  -0.831 0.407403
---
Signif. codes:  0 '***' 0.001 '**' 0.01 '*' 0.05 '.' 0.1 ' ' 1

Residual standard error: 1.889 on 159 degrees of freedom
Multiple R-squared: 0.976,       Adjusted R-squared: 0.9745
F-statistic: 646.8 on 10 and 159 DF,  p-value: < 2.2e-16
```

Incorporating the linear variables into the model, for the training part we get only a negligible improvement of the coefficient of determination R^2.

Again we can use the functions from `library(leaps)` to search for "optimal" models.

Input
```
library(leaps)
l1 <- leaps(x = fat[, c(6:15, 20:29)], y = fat$vol)
```

Of course statistical analysis cannot substitute for subject knowledge. In a real consulting situation, it is a difficult task to communicate to the client that, for example, giving an optimal estimate of the volume is something different from giving an optimal estimate of 1/volume and then inverting it. It is up to the expert in the subject to know (or at least to decide) which is the relevant parameter scale. Does body fat percentage really matter? Or is the average tissue density the parameter of interest?

If we want to model the proportion of body fat directly, not the volume, we can construct the corresponding auxiliary variables.

Input

```
fat$neckvolf <- fat$neckvol / fat$weightkg
fat$chestvolf <- fat$chestvol / fat$weightkg
fat$abdomenvolf <- fat$abdomenvol / fat$weightkg
fat$hipvolf <- fat$hipvol / fat$weightkg
fat$thighvolf <- fat$thighvol / fat$weightkg
fat$kneevolf <- fat$kneevol / fat$weightkg
fat$anklevolf <- fat$anklevol / fat$weightkg
fat$bicepvolf <- fat$bicepvol / fat$weightkg
fat$forearmvolf <- fat$forearmvol / fat$weightkg
fat$wristvolf <- fat$wristvol / fat$weightkg
```

We begin with a simple model. We use only one variable (`abdomenvolf`) from the group of variables describing the torso, and one variable (`wristvolf`) from the higher principal components. With this simple model, we achieve about the same quality as previously with the full set of variables.

Input

```
lm.volf <- lm(body.fat ~  abdomenvolf  + wristvolf,
    data = fat,
    subset = fat$train)
summary(lm.volf)
```

Output

```
Call:
lm(formula = body.fat ~ abdomenvolf + wristvolf, data = fat,
    subset = fat$train)

Residuals:
    Min      1Q  Median      3Q     Max
-9.4758 -3.0952  0.1740  3.2076  8.5131

Coefficients:
              Estimate Std. Error t value Pr(>|t|)
(Intercept) -2.3046435  5.8696465  -0.393    0.695
abdomenvolf  0.0024999  0.0001774  14.095  < 2e-16 ***
wristvolf   -0.0348388  0.0048811  -7.138 2.77e-11 ***
---
Signif. codes:  0 '***' 0.001 '**' 0.01 '*' 0.05 '.' 0.1 ' ' 1

Residual standard error: 4.161 on 167 degrees of freedom
Multiple R-squared: 0.7025,     Adjusted R-squared: 0.699
F-statistic: 197.2 on 2 and 167 DF,  p-value: < 2.2e-16
```

So the recommendation would be: if you do not have a bath tube with you to determine the volume (or the density), measure weight, height and the abdomen and wrist circumference. Do not use a linear model in the original data: transform the circumference to volumes.

Exercise 4.12	
*	Extend the variables by other volume-related variables in the model given above. Do you gain precision?
**	Try to include the variable `age` in the model. How exactly do you include `age` in the model?
**	The function *mvr()* in *library(pls)* [54] is available to perform a regression based on principal components. Use this function for regression. What is the difference between this estimation and the usual least squares regression?

For model construction, we used only the training part of the data. The quality of the model derived now can be checked using the evaluation part. This can be done using function *predict.lm()*, which applies a model estimated with *lm()* to a new data set with analogous structure, for example:

Input
```
fat.eval <- fat[fat$train == FALSE, ]
pred <- predict.lm(lm.volf, fat.eval, se.fit = TRUE)
```

Exercise 4.13	
*	Estimate the precision of the model using the evaluation part of the data.
*	Carry out a regression diagnostics of the model derived, using the evaluation part of the data.

4.8 High Dimensions

In small dimensions, we can analyse many problems completely. Higher dimensions often require us to design special strategies for analysis. Formal applications of standard methods soon meet their limitations.

Higher dimensions, such as dimension 10 through 100, are common in many application fields. But even problems in really high dimensions occur day to day. Dimension is a question of modelling, not a mere question of the problem. Digital video (DV PAL) for example records images in a format of 720×576. A single image with three colours hence gives a vector in $720 \times 576 \times 3 = 1244160$-dimensional space. A second of video has 25 of these frames. If we have to work with image data, it is our choice to see image processing as a problem in dimension $d = 1244160$, or as a sequence of 1244160 (not independent!) observations in dimension $d = 1$.

Going from dimension $d = 1244160$ to $d = 1$ we shift information that is implicit in the dimensions to structural information. Of course, we have to pay for it: now the structural information has to be modelled.

A comment: practical application takes a middle course. The image is decomposed into blocks, for example of size 64×64. Pixels within a block are handled simultaneously; the blocks are handled sequentially. You can see these blocks the next time your digital TV has a glitch.

In high-dimensional problems, the statistics is often not apparent at all. It is hidden in the hardware as an "imbedded system".

Figure 4.12 is an example from an analysis with R for a high-dimensional data set of cDNA micro array data ([44]). A single observation in this data set consists of measurements on 4227 probes, each with four partial recordings for two colours (red and green) and two attributes (foreground fg and background bg), giving (*fg.green*, *fg.red*, *bg.green*, *bg.red*). This means one data point has dimensions 4×4227. This information is physically related to positions on the slide, the spots. The essential function that is used for visualisation here is *image()*. It is used here to represent a variable *z* using a colour table against two coordinates *x, y*, which give the position of a spot on a slide. The image shows one observation. The four channels for the partial recordings are put on different displays side by side, arranged by spot position. A unified yellow/blue colour table is chosen, which is recognisable even for most colour perception deficiencies. We used a similar strategy earlier in Section 2.4.2 (page 88).

In this part of the analysis, quality control was the main issue. A starting point is to identify "dirty" spots, that is, spots where the background intensity exceeds the foreground intensity. These are identified as red spots, and their distribution is represented in special plots. The other critical information is on the adjustment of the channels. This information is represented in the bottom line. In this series of experiments, there should be a small fraction of differentially activated genes. So the distribution of the red intensity should be approximatively that of the green intensity. For this specific chip, we see that the estimated densities for the red and green dies clearly differ, as shown in the bottom left plot. So this particular chip has a quality problem. You can see a video of the production process in `<http://www.statlab.uni-heidelberg.de/projects/genex/cdna.mpg>` and with the help of the diagnostics you should be able to spot the critical points.

What we have illustrated here is an example of an analysis of dimensions 4×4227. It is feasible, but of course this is not covered by standard methods.

The colours code the result of a pre-analysis — the red points mark problem areas on the cDNA-Chip. In this case, the pattern of the highlighted selection could identify a specific problem in the production process: the slides are incubated in a small reaction chamber, and small differences in the pressure applied when closing the chamber could lead to a spatial inhomogeneity in the reaction.

Theme-oriented surveys about R packages, in particular for multivariate problems, are to be found at `<http://cran.at.r-project.org/src/contrib/Views/>`.

4.9 Statistical Summary

The analysis of multivariate data can only be touched on in this context. Multivariate problems already occur implicitly for regression problems (see Chapter 2). For simple regression problems, the multivariate aspects only referred to deterministic parameters. In the general case we have to analyse a multivariate statistical distribution. We finish our introduction here and reserve further discussion for future lectures.

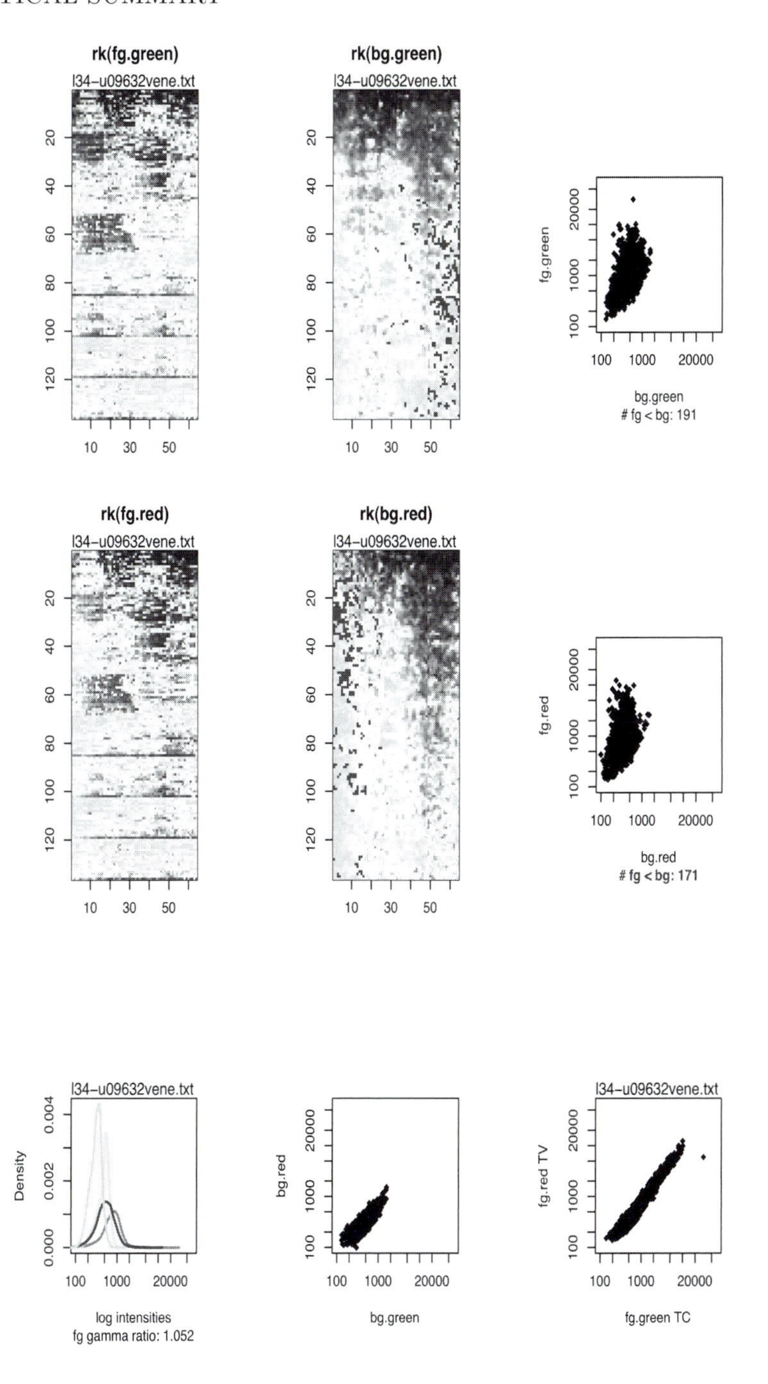

Figure 4.12 *A single* 4227×4*-dimensional observation from a micro array experiment. See Colour Figure 12.*

R as a Programming Language and Environment

R is an interpreted expression language. Expressions are composed of objects and operators.

A.1 Help and Information

Some R functions such as *library()* or *data()* serve a dual purpose. With minimal arguments, they provide help and information. With specific arguments, they give access to certain components.

R ***Help***	
help()	information about an object/a function. *Example:* *help(help)*
help.start()	starts browser access to R's online documentation. The reference section includes a search engine to search for keywords, function and data names and text in help page titles.
args()	shows arguments of a function.
example()	executes examples, if available. *Example:* *example(plot)*
help.search()	searches for information about an object/a function.
RSiteSearch()	searches for keywords or phrases in the R-help archives or documentation.
apropos()	locates by keyword.
demo()	executes demos for a topic area. *Example:* *demo(graphics)* *demo()* lists all topic areas that provide a demo.
library()	gives information about libraries. *Example:* *library()* gives a list of all libraries.

(cont.)→

R ***Help*** (cont.)	
	`library(help=⟨package⟩)` gives information about a package. *Example:* `library(help="stats")` gives information about the basic statistics package.
`data()`	gives information about data sets. *Example:* `data()` lists available data sets.
`vignette()`	lists or views vignette information about a topic. `vignette(all = TRUE)` lists vignettes from all installed packages. *Example:* `vignette("grid")` shows a vignette for the grid graphics.

See also Appendix A.6 "Object Inspection" (page A-200) and Appendix A.7 "System Inspection" (page A-201).

A.2 Names and Search Paths

Objects are identified by names. By the name objects are searched in a search path, a chain of search environments. The search path in effect can be inspected with `search()`.

R *Search Paths*	
`search()`	lists the search areas in effect, beginning with `.GlobalEnv` down to the base package `package:base`. *Example:* `search()`
`searchpaths()`	lists the access paths for the search areas in effect. *Example:* `searchpaths()`
`objects()`	lists the objects in a search path. *Examples:* `objects()` `objects("package:base")`
`ls()`	lists the objects in a search path. *Examples:* `ls()` `ls("package:base")`
`ls.str()`	lists the objects and their structure in a search path. *Examples:* `ls.str()` `lsf.str("package:base")`
`find()`	locates by keyword. Also finds overlaid entries. *Syntax:* `find(what, mode = "any", numeric = FALSE, simple.words = TRUE)`
`apropos()`	locates by keyword. Also finds overlaid entries. *Syntax:* `apropos(what, where = FALSE, ignore.case = TRUE, mode = "any")`

Functions can be nested. This may occur at definition time as well as at execution time. This requires an extension of the search paths. The dynamic identification of objects uses environments to resolve local or global variables in functions.

R *Search Paths (cont.)*	
`environment()`	current environments. *Example:* `environment()`
`sys.parent()`	preceding environments. *Example:* `sys.parent(1)`

A.3 Administration and Customisation

objects() *ls()*	lists the objects in the current search path.
rm()	removes indicated objects. *Syntax:* `rm(⟨object list⟩)`

R offers a series of possibilities to configure the system so that certain commands are executed upon start or termination. When starting, the files *.Rprofile* and *.RData* are read and executed if available. Details can be system specific. The appropriate information is given by:

```
help(Startup)
```

Various parts of the system keep global information and can be configured by setting options and parameters.

Some System Components with Global State	
basic system	see *help(options)*.
random numbers	see Appendix A.21 (page A-228).
basic graphics	see *help(par)*.
lattice graphics	see *help(lattice.options)*.

For information on how to configure memory available for data storage, see:

```
help(Memory)
```

See also Appendix A.7 "System Inspection" (page A-201).

A.4 Basic Data Types

Basic R ***Data Types***	

`numeric`	`real` or `integer`. In R: real numbers are always in double precision. Single precision is supported for external call to other languages with .C or .FORTRAN. Functions `mode()` and `typeof()` can show the storage modus (single, double ...), depending on the implementations.
	Examples: `1.0`
	`2`
	`3.14E0`

`complex`	complex, in Cartesian coordinates. *Example:* `1.0+0i`
`logical`	TRUE, FALSE. In R, T and F are predefined variables provided as an alternative. In S-Plus, T and F are basic objects.
`character`	character strings. Delimiter are alternatively " or '. *Example:* `"T"`, `'klm'`
`list`	general list structure. List elements can be of different types. *Example:* `list(1:10, "Hello")`
`function`	R function. *Example:* `sin`
`NULL`	special case: empty object. *Example:* `NULL`

`is.⟨type⟩()` tests for a type, `as.⟨type⟩()` converts to a type.

In addition to TRUE and FALSE there are three special values for exceptional situations:

Special Constants	
`TRUE`	alternative: *T*. Type: logical.
`FALSE`	alternative: *F*. Type: logical.
`NA`	"not available". Type: logical. NA is different from TRUE and FALSE.
`NaN`	"not a valid numeric value". Implementation dependent. Should follow the IEEE Standard 754. Type: numeric. *Example:* `0/0`

(cont.)→

Special Constants (cont.)	
Inf	infinite. Implementation dependent. Should follow the IEEE Standard 754. Type: numeric. *Example:* `1/0`

Test Functions	
is.na()	returns *TRUE* if the argument has the value *NA* or *NaN*.
na.omit()	returns an object with the cases containing *NA* removed.
na.fail()	returns its argument if no the case contains *NA*; signals an error message otherwise.
is.nan()	returns *TRUE* if the argument has the value *NaN*.
is.inf()	returns *TRUE* if the argument has the value *Inf* or *-Inf*.

A.5 Output for Objects

The object attributes and content can be queried or displayed using output routines. The output routines generally are ***polymorphic***, that is they come with variants adapted to the given object type. To list all available methods for an generic function, or all methods for a class, use `methods()`, for example `methods(print)`.

R *Inspection*	
`print()`	standard output.
`cat()`	outputs the objects, concatenating the representations. `cat()` is useful for producing output in user-defined functions, with minimal formatting.
`format()`	formats an R object for pretty printing.
`structure()`	output, optional with attributes.
`summary()`	standard output as summary, in particular for model fits.
`plot()`	standard graphic output.

For converting tables to a HTML or LaTeX format, `library(xtable)` [7] is available.

Output of objects to files is discaed in Appendix A.15 "Input and Output to Data Streams" (page A-215).

A.6 Object Inspection

Objects have two implicit attributes that can be queried with `mode()` and `length()`. The function `typeof()` gives the (internal) storage modus of an object.

A `class` attribute gives the class of an object.

The following table summarises the most important information possibilities about objects.

Object Inspection	
`str()`	shows the internal structure of an object in compact form. *Syntax:* `str(⟨object⟩)`
`structure()`	shows the internal structure of an object. Attributes for the display can be passed as parameters. *Example:* `structure(1:6, dim = 2:3)` *Syntax:* `structure(⟨object⟩, ...)`
`class()`	object class. For object clases defined in newer R versions, the class is stored as an attribute. For vintage object classes, the class is determined implicitly by type and other attributes.
`mode()`	mode (type) of an object.
`storage.mode()`	storage mode of an object.
`typeof()`	mode of an object. May be different from the storage mode. Depending on the implementation a numerical variable, for example, can be stored in double precision (the default) or in single precision.
`length()`	length = number of elements.
`attributes()`	reads/sets attributes of an object, such as names, dimensions, classes.
`names()`	names attribute for elements of an object, for example, a vector. *Syntax:* `names(⟨obj⟩)` gives the `names` attribute of `⟨obj⟩`. `names(⟨obj⟩)<-⟨charvec⟩` sets the `names` attribute. *Example:* `x<-values` `names(x)<- ⟨charvec⟩`

A.7 System Inspection

The following table summarises the most important information possibilities about the general system environment. When used with an argument, these functions generally serve specific purposes, such as setting parameters and options. When used with an empty argument list, they provide inspection.

System Inspection	
`search()`	current search path.
`ls()`	objects in current or selected search path.
`methods()`	generic methods: *Syntax:* `methods(⟨fun⟩)` shows specialised functions for ⟨fun⟩, `methods(class = ⟨c⟩)` the class-specific functions for class ⟨c⟩. *Examples:* `methods(plot)` `methods(class = lm)`
`data()`	accessible data.
`library()`	accessible packages.
`help()`	general help system.
`options()`	global options.
`par()`	parameter settings for the graphics system.
`capabilities()`	reports availability of optional features.

The options of the `lattice` systems can be controlled with `trellis.par.set()` resp. `lattice.options()`.

R is anchored in the host operating system. Some variables such as access paths, encoding, etc. are imported from there.

System Environment	
`getwd()`	gets current working directory.
`setwd()`	sets current working directory.
`dir()`	lists files in the current working directory.
`system()`	calls system functions.

A.8 Complex Data Types

The interpretation of basic types or derived types can be specified by one or more `class` attributes. Polymorphic functions such as `print` or `plot` evaluate this attribute and call a variant for this class if available (see Section 2.6.5 (page 103)).

For the storage of dates and times, special classes are provided. For more information on these data types see

```
help(DateTimeClasses)
```

and Appendix A.15 (page A-215).

R is vector based. Individual constants or values just are vectors with the special length 1. They do not get a special treatment.

Compound Data Types		
Vectors	basic R data types.	
Matrices	vectors with two-dimensional layout.	
	See also	Appendix A.9 "Data Manipulations" (page A-205).
Arrays	vectors with higher-dimensional layout.	
	`dim()`	defines a dimension attribute.
	Example:	`x < -runif(100)` `dim(x) < - c(5, 5, 4)`
	`array()`	generates a new vector with specified dimension structure.
	Example:	`z < - array(0, c(4, 3, 2))`
	See also	Appendix A.9 "Data Manipulations" (page A-205).
Factors	special case for categorical data.	
	`factor()`	converts a numeric vector into a factor.
	See also	Section 2.2.1.
	`ordered()`	converts a vector into a factor with ordered levels. This is a shortcut for `factor(x, ..., ordered = TRUE)`.
	`levels()`	returns the levels of a factor.

	Example:	`x <- c("a", "b", "a", "c", "a" )`
		`xf <- factor(x)`
		`levels(xf)`
		results in
		`[ 1 ] "a" "b" "c"`

	`tapply()`	applies a function separately for all levels of factors in a list.

(cont.)→

Compound Data Types (cont.)	
Lists	analogous to vectors, with elements of possibly different types. *list()* generates a list. *Syntax:* list(⟨components⟩) *[[ ]]* access to components of a list by index. ⟨list$component⟩ access by names. *Example:* l <- list(name = "xyz", age = 22, fak = "math") > l[[2]] 22 >l $age 22
Data Frames	**data frames** analogous to arrays resp. lists, with column-wise uniform type and uniform column length. *data.frame()* analogous to *list()*, but restrictions have to be satisfied. *attach()* attaches a database to the current search list. For access to components the component name will be sufficient. *detach()*

A.9 Accessing Components

The length of vectors is a dynamic attribute. It is extended or shortened as needed. In particular, an implicit "recycling rule" applies: if a vector does not have the length necessary for some operation, it is repeated periodically up to the length required.

Vector components can be accessed by index. The indices can be specified explicitly or in the form of an expression rule.

Accessing Components	
`x[⟨indices⟩]`	indicated components of x. *Example:* `x[1:3]`
`x[-⟨indices⟩]`	x omitting indicated components. *Example:* `x[-3]` x omitting the 3. component.
`x[⟨condition⟩]`	components of x, for which the ⟨`condition`⟩ holds. *Example:* `x[x<0.5]`
`which()`	give the indices of a logical object, allowing for array indices.
`subset()`	is a polymorphic function and returns subsets of vectors, matrices or data frames by specified conditions.

Vectors (and other objects) can be mapped to higher-dimensional constructs. The layout is described by a additional `dim` attribute. By convention the imbedding goes by column, that is, the first index varies first (FORTRAN convention). Operators and functions can evaluate the dimension attribute.

R *Index Access*	
`dim()`	gets or sets dimensions of an object. *Example:* `x <- 1:12; dim (x) <- c(3, 4)`
`dimnames()`	gets or sets names for the dimensions of an object.
`nrow()`	gives the number of rows = dimension 1.
`ncol()`	gives the number of columns = dimension 2.
`matrix()`	generates a matrix with given specifications. *Syntax:* `matrix(data = NA, nrow = 1, ncol = 1, byrow = FALSE, dimnames = NULL)` *See also* Example 1.11 (page 23)
`array()`	generates a possibly higher-dimensional matrix. *Example:* `array(x, dim = length(x), dimnames = NULL)`

`NCOL()` and `NROW()` are variants treating a vector as a one-column resp. as a one-row matrix.

R *Iterators*	
`apply()`	applies a function to the rows or columns of a matrix. *Syntax:* `apply(x, MARGIN, FUNCTION, ...)` MARGIN = 1: rows, MARGIN = 2: columns *See also* Example 1.11 (page 23).
`lapply()`	applies a function to the elements of a list. *Syntax:* `lapply(X, FUN, ...)`
`sapply()`	applies a function to the elements of a list, of a vector or a matrix. If possible, dimension names are carried over. *Syntax:* `sapply(X, FUN, ..., simplify = TRUE, USE.NAMES = TRUE)`
`mapply()`	applies a function to multiple list or vector arguments. *Syntax:* `mapply(FUN, ..., MoreArgs = NULL,simplify = TRUE, USE.NAMES = TRUE)`
`Vectorize()`	returns a new function that acts as if `mapply` was called. This can be used as a stepping stone to make a function vectorized. *Syntax:* `Vectorize(FUN, vectorize.args = arg.names,simplify = TRUE, USE.NAMES = TRUE)`
`tapply()`	applies a function to components of an object depending on a list of controlling factors.
`by()`	object-oriented variant of `tapply`. *Syntax:* `by(data, INDICES, FUN, ...)`
`aggregate()`	calculates statistics for subsets. *Syntax:* `aggregate(x, ...)`
`replicate()`	evaluates an expression repeatedly (for example, with generating random numbers for simulation). *Syntax:* `replicate(n, expr, simplify = TRUE)`
`outer()`	generates a matrix with all pair-wise combinations from two vectors, and applies a function to each pair. *Syntax:* `outer(vec1, vec2, FUNCTION, ...)`

A.10 Data Manipulation

Array Access	
`cbind()`	combines by columns.
`rbind()`	combines by rows.
`split()`	splits a vector by factors.
`table()`	generates a table of counts.
`prop.table()`	expresses table entries as fraction of marginal table, i.e., gives relative counts.
`t()`	transposes rows and columns. *Syntax:* `t(x)`
`aperm()`	generalised permutation. *Syntax:* `aperm(x, perm)` where `perm` is a permutation of the indices of `x`.

Transformations	
`duplicated()`	checks for duplicate or multiple values.
`unique()`	generates a vector without multiple values.
`match()`	gives first position of a value in a vector.
`pmatch()`	partial matching

Character String Transformations	
`casefold()`	translates characters, in particular from upper - to lowercase or vice versa.
`tolower()`	translates to lowercase.
`toupper()`	translates to uppercase.
`chartr()`	translates characters in a character vector.
`substr()`	extracts or replaces substrings in a character vector.
`substring()`	extracts or replaces substrings in a text (respets encoding and other attributes)
`paste()`	concatenates vectors after converting to character. See also `cat()`.
`strsplit()`	splits the elements of a character vector into substrings.

(cont.)→

Character String Transformations (cont.)	
`grep()`	pattern matching.
`gsub()`	pattern substitution, by regular patterns.
`abbreviate()`	abbreviates strings.

Transformations	
`table()`	generates a table of counts.
`expand.grid()`	generates a data frame with all combinations of the factors given.
`gl()`	generates factors by specifying the pattern of their levels.
`reshape()`	converts between a cross classification table (column per variable) and a long table (variables in rows, with additional indicator column).
`merge()`	merges data frames. See `help(merge)` for examples. `merge()` supports various versions of data base `join` operations.

Vector Manipulation	
`seq()`	generates a sequence.
`stack()`	concatenates multiple vectors from a data frame or list into a single vector and generates a factor indicating the source of each item. *Syntax:* `stack(x, ...)`
`unstack()`	splits a vector by an indicator variable, i.e., reverses the operation of `stack()`. *Syntax:* `unstack(x, ...)`
`split()`	splits a vector into the groups defined by a factor. *Syntax:* `split(x, f, drop = FALSE,...)`
`unsplit()`	combines components to a vector, i.e., reverses `split()`. *Syntax:* `unsplit(value, f, drop = FALSE)`
`cut()`	converts a numeric to factor. `cut()` divides the range of a vector into intervals and creates a factor indicating the interval for each value. *Syntax:* `cut(x, ...)`

A.11 Operators

Expressions in R can be composed of objects and operators. The following table of operators is ordered by precedence (highest rank on top). See `help(Syntax)`.

Basic R operators	
`$`	select component by name. *Example:* `list$item`
`[ [[`	indexing, access to elements. *Example:* `x[i]`
`^`	exponentiation (right to left). *Example:* `x^3`
`-`	unary minus.
`:`	sequence generation. *Examples:* `1:5` `5:1`
`%⟨name⟩%`	special operators. Can also be user defined. *Examples:* `"%deg2%"<-function(a, b) a + b^2` `2 %deg2% 4`
`* /`	multiplication, division.
`+ -`	addition, subtraction.
`< > < = > = ==` `!=`	comparison operators.
`!`	negation.
& | && ||	and, or . &&, || are “shortcut” operators.
`<- ->`	assignment.

If operands do not have the same length, the shorter operand is repeated cyclically.

Operators of the form %⟨name⟩% can be defined by the user. The definition follows the rules for function definitions.

Expressions can be written as a sequence with separating semicolons. Expression groups can be combined by enclosing braces {...}.

A.12 Functions

Functions are special objects. Functions can return objects as results.

R *Function Declarations*	

Declarations	`function ( ⟨formal argument list⟩ )`
	`⟨expression⟩`
	Example: `fak <- function(n) prod(1:n)`

Formal argument	`⟨argument name⟩` `⟨argument name⟩ = ⟨default value⟩`
Formal argument list	list of formal argument, separated by commas. *Examples:* `n, mean = 0, sd = 1`
...	variable argument list. Variable argument lists can be propagated to imbedded functions. *Example:* `mean.of.all <- function (...)mean(c(...))`
Function result	`return ⟨value⟩` stops function evaluation and returns value.
	`⟨value⟩` as last expression in a function declaration: returns value.
Assignments	In general, assignments operate only on local copies of variables. Assignments done within a function are temporary. They are lost after exit from the function. The assignment with `<<-`, however, looks for the target in the complete search chain. It can be used if global and permanent assignments are intended within a function. *Syntax:* `⟨Variable⟩<<-⟨value⟩`

R *Function Call*	
Function call	`⟨name⟩(⟨Supplied (actual) argument list⟩)` *Example:* `fak(3)`
Supplied argument list	Values are matched by position. Deviating from this, names can be used to control the matching. Initial parts of the names suffice (exception: after a variable argument list, names must be given completely). Function `missing()` can be used to check, whether a corresponding actual argument is missing for a formal argument. *Syntax:* `⟨list of values⟩` `⟨argument name⟩ = ⟨values⟩` *Example:* `rnorm(10, sd = 2)`

Arguments for functions are passed by value. This helps consistency, but involves overhead for memory management and copying. If this overhead needs to be avoided, the information provided by *environment()* allows direct access to variables. Techniques to use this are described in [12].

Special case: Functions with names of the form `xxx<-` extend the assignment function. ***Example:***

Input

```
"inc<-" <-function (x, value) x+value
x <- 10
inc(x) <- 3
x
```

Output

```
[1] 13
```

In R assignment functions the value argument **must** be called "value".

A.13 Debugging and Profiling

R provides a collection of tools for identification of errors. These are particularly helpful in connection with functions. *browser()* can be used to switch to a browser mode. In this mode, the usual R instructions can be used. Besides this, there is a small number of special instructions. With *debug()*, the browser mode is activated automatically upon entry to a function. The browser mode is marked by a special prompt *Browse[xx]*>.

⟨**return**⟩	Goes to the next instruction, if the function is under control of *debug*. Continuous with the expression evaluation if *browser* has been called directly.
n	Goes to the next instruction (also if *browser* has been called directly).
cont	Continuous with the expression evaluation.
c	Short for *cont*. Continues the expression evaluation.
where	Shows call nesting.
Q	Stops execution and jumps back to base state.

Debug Help	
browser()	suspends execution and enters the browser mode. *Syntax:* *browser()*
recover()	shows a list of the current call hierarchy. An entry from this list can be chosen for inspection by *browser()*. With *c* you leave the *browser* and return to *recover*. With *0* you leave *recover()* *Syntax:* *recover()* *Hint:* With *options(error = recover)*, error handling for a function is directed to call *browser()* automatically in case of an error.
debug()	marks a function for debugger control. On subsequent calls to the function, the debugger is activated and switches to browser mode. *Syntax:* *debug(*⟨**function**⟩*)*
undebug()	cancels debugger control for a function. *Syntax:* *undebug(*⟨**function**⟩*)*
trace()	marks a function for trace control. On subsequent calls to the function, the call is signalled together with its arguments. *Syntax:* *trace(*⟨**function**⟩*)*
untrace()	cancels trace control for a function. *Syntax:* *untrace(*⟨**function**⟩*)*

(cont.)→

Debug Help (cont.)	
traceback()	in case of error inside of a function the current calling stack is stored in a variable *.Traceback*. *traceback()* evaluates this variable and displays its content. *Syntax:* *traceback()*
try()	Calls a function. Allows for user-defined error handling. *Syntax:* *traceback(*⟨expression⟩*)*

To measure execution time in selected code ranges, R provides a "profiling". This is only available if R has been compiled with the appropriate options. The options installed at compiling can be queried using *capabilities()*. See Appendix A.7 "System Inspection" (page A-201).

Profiling Support	
system.time()	returns the execution time of an expression. This function is available in all implementations. *Syntax:* *system.time(*⟨expr⟩, ⟨gcFirst⟩*)*
Rprof()	records active functions periodically. This function is only available if R has been compiled for "profiling". With *memory.profiling = TRUE*, in addition to the timing the memory usage is recorded periodically. This option is only available if R has been compiled correspondingly. *Syntax:* *Rprof(filename = "Rprof.out", append = FALSE, interval = 0.02, memory.profiling = FALSE)*
Rprofmem()	records memory requirements on demand. This function is only available if R has been compiled for "memory profiling". *Syntax:* *Rprofmem(filename = "Rprofmem.out", append = FALSE, threshold = 0)*
summaryRprof()	summarises the output of *Rprof()* and reports the timing by function. *Syntax:* *summaryRprof(filename = "Rprof.out", chunksize = 5000, memory = c("none", "both", "tseries", "stats"), index = 2, diff = TRUE, exclude = NULL)*

A.14 Control Structures

R *Control Structures*	
if	conditional execution. *Syntax:* `if (⟨log. expression 1⟩) ⟨expression2⟩` The logical expression1 may return only one logical value. For vector-oriented access use *ifelse*. *Syntax:* `if (⟨log. expression1⟩) ⟨expression2⟩ else ⟨expression3⟩`
ifelse	element wise conditional execution. *Syntax:* `ifelse(⟨log. expression1⟩, ⟨expression2⟩, ⟨expression3⟩)` evaluates the logical expression1 elementwise on a vector, and returns expression2 if the evaluation gives true, else expression3. *Example:* `trimmedX <- ifelse (abs(X)<2, x, sign(X)*2)`

switch	evaluates an expression and executes an instruction based on the result.
	Syntax: `switch(⟨expression1⟩, ...)`
	expression1 must return a numeric value or a character string. ... is an explicit list of alternative actions.
	Example: `centre <- function (x , type) { switch(type,`
	`mean = mean(x),`
	`median = median(x),`
	`trimmed = mean(x, trim = .1)}`

for	iteration (loop). *Syntax:* `for (⟨name⟩ in ⟨expression1⟩) ⟨expression2⟩`

repeat	iteration. Must be terminated explicitly, for example with *break*.
	Syntax: `repeat ⟨expression⟩`
	Example: `pars<-init`
	`repeat { res<- get.resid (data, pars)`
	`if (converged(res) ) break`
	`pars<-new.fit (data, pars)}`

(cont.)→

R *Control Structures* (cont.)	

`while`	conditional repetitions.
	Syntax: `while (⟨log. expression⟩) ⟨expression⟩`
	Example: `pars<-init; res <- get.resid (data, pars)`
	`while (!converged(res)) { pars<-`
	`new.fit(data, pars)`
	`res<- get.resid}`

`break`	terminates the current loop and exits.
`next`	terminates the current loop cycle and advances to next cycle.

Note: In R loops should be avoided if possible in favor of more efficient language constructs (see [25]).

A.15 Input and Output to Data Streams; External Data

R *Input/ Output*	
`write()`	writes data to a file. *Syntax:* `write(val, file)` *Example:* `write(x, file = "data")`
`source()`	executes the R instruction from the file indicated. *Syntax:* `source("⟨file name⟩")` *Example:* `source("cmnds.R")`
`Sweave()`	executes the R instruction from the file indicated and entangles embeddded text. Sweave can be used for automatiic report generation. *Syntax:* `Sweave("⟨file name⟩", ...)`
`sink()`	redirects output in the file specified. *Syntax:* `sink("⟨file name⟩")` *Example:* `sink()` redirects the output back to the console.
`dump()`	writes the commands defining an object. The object can be regenerated from this output using `source()`. *Syntax:* `dump(list, file = "⟨dumpdata.R⟩", append = FALSE)`

R can access data from local files indicated by a usual file path or from remote files accessed by an URL reference. On most systems, direct access to a clipboard is available as well. More system-specific information is available using `help(connections)`.

To edit or enter data, R provides `edit()`. This is a polymorphic function For the special case of matrix-like data, `data.entry()` is provided, using a spreadsheet model.

For exchange, the data formats have to be harmonised between all parties. For import from data bases or other systems, several packages are available, for example `library(stataread)` for Stata, `foreign` for SAS, Minitab and SPSS, `library(RODBC)` for SQL. For more information, see the manual "Data Import/Export" [32].

Within R, prepared data are usually provided as `data frames`. If additional objects such as functions or parameters are necessary, they can be made accessible in bundled form as packages. See Appendix A.16 (page A-218).

For the exchange from R to R, a special exchange format can be used. Files in this format can be generated with `save()` and conventionally have the name suffix `.Rda`. These files can be loaded again using `load()`.

A general purpose function to load data us `data()`. Depending on the suffix of the input file name, `data()` branches for several special cases. Besides `.Rda` usual suffixes for data input files are `.tab` or `.txt`. The online help function `help(data)` gives additional information.

Data Input/Output for R	
save()	stores data in an external file. *Syntax:* `save(⟨names of the objects to be stored⟩, file = ⟨file name⟩, ...)`
save.image()	is a short-cut and stores data of the workspace in an external file.
load()	loads data from an external file. *Syntax:* `load(file = ⟨file name⟩, ...)`
data()	loads data. *data()* can handle various file formats, if the access paths and filenames follow the R conventions. *Syntax:* `data(... , list = character(0), package = c(.packages(), .Autoloaded), lib.loc = .lib.loc)` *Example:* `data(crimes)   # loads the data set 'crimes'`

For the flexible exchange with other programs in general text-based files are provided. Some conventions can make exchange easier:

- in table form
- only ASCII characters (for example, no umlaut!)
- variables arranged in columns
- columns separated by tabulator stops
- possibly a column header in row 1
- possibly a row label in column 1

For reading the function *read.table()* is provided, and for writing, there is *write.table()*. Besides *read.table()* there are several variants that are adapted to usual data formats. These are documented under *help(read.table)*.

Input and Output of Data for Exchange	
read.table()	reads data tables. *Syntax:* `read.table(file, header = FALSE, sep = "\t", ...)` *Examples:* `read.table(⟨file name⟩, header = TRUE, sep = "\t")` headers in row 1, row labels in column 1 `read.table(⟨file name⟩, header = TRUE, sep = '\t')` now row number, headers in row 1,

(cont.)→

Input and Output of Data for Exchange (cont.)	

`write.table()`	writes data table.
	Syntax: `write.table(file, header = FALSE, sep = '\t', ...)`
	Examples: `write.table(`⟨`data frame`⟩, ⟨`file name`⟩`, header = TRUE,`
	`sep = '\t')`
	headers in row 1, row labels in column 1
	
	`write.table(`⟨`data frame`⟩, ⟨`file name`⟩,
	`header = TRUE, sep = '\t')`
	now row number, headers in row 1.

`read.csv()`	reads comma-separated data tables.
`write.csv()`	writes comma-separated data tables.
`read.csv2()`	reads semicolon-separated data tables, using a comma as decimal separator.
`write.csv2()`	writes semicolon-separated data tables, using a comma as decimal separator.

By default, `read.table()` converts data to `factor` variables if possible. This behaviour can be modified with the argument `as.is` when calling of `read.table()`. This modification is, for example, necessary to read date and time information as for example in the following example from [13]:

```
# date col in all numeric format yyyymmdd
df <- read.table("laketemp.txt", header = TRUE)
as.Date(as.character(df$date), "%Y-%m-%d")
# first two cols in format mm/dd/yy hh:mm:ss
# Note as.is = in read.table to force character
library("chron")
df <- read.table("oxygen.txt", header = TRUE,  as.is = 1:2)
chron(df$date, df$time)
```

For sequential reading, `scan()` is provided. Files with data in fixed format (by character columns) can be read with `read.fwf()`.

A.16 Libraries, Packages

External information can be stored in (text) files and packages. In general, additional functions are provided as packages. Packages may be installed as part of the basic installation or installed by the user. Once packages are installed, they are loaded with

```
library()
```

when needed. Data sets contained in the package are then included in the search path and can be listed using `data()` without arguments:

```
data()
```

Example:

```
library(nls)
data()
data(Puromycin)
```

If you use R packages, please treat them as you would treat any other scientific source of information. Credit should be given where credit is due, and proper citations should be included. The function `citation()` gives the bibliographic information to use.

Package Utilities	
`install.packages()`	installs add-on package in ⟨`lib`⟩, downloading it from the archive *CRAN* or from specified archives. *Syntax:* `install.packages(pkgs, lib, CRAN = getOption("CRAN"), ...)` *Example:* `install.packages("mypackage.tgz", repos=NULL` installs package from a local file.
`library()`	loads an installed add-on package into the current workspace. *Syntax:* `library(package, ...)` *See also* Section 1.5.6 "Packages" (page 54).
`require()`	tries to load an add-on package; gives warning on error. *Syntax:* `require(package, ...)`
`detach()`	releases an add-on package and removes it from the search path. *Syntax:* `detach(⟨name⟩)`
`package.manager()`	if implemented, interface for management of installed packages. *Syntax:* `package.manager()`
`package.skeleton()`	creates a skeleton for a new package. *Syntax:* `package.skeleton(name = "⟨anRpackage⟩", list, ...)`
`citation()`	gives bibliographic information for citing a package. *Syntax:* `citation(⟨package name⟩, lib.loc = NULL)`

For Unix/Linux/Mac OS X, the main tools are available as commands:

```
R CMD check <directory> # checks a directory for compliance with the R conventions
R CMD build <directory> # generates an R package
```

Detailed information for building R packages is in “Writing R Extensions” ([35]).

A.17 Mathematical Operators and Functions; Linear Algebra

For basic arithmetic operators, see `help(Arithmetic)`. For trigonometric functions, information is available using `help(Trig)`. For special mathematical functions, including `beta()`, `factorial()`, `choose()`, see `help(Special)`.

For linear algebra, the most important functions are widely standardised and implemented in C libraries such as BLAS/ATLAS and LAPACK. R makes use of these libraries and provides an interface to the most important functions.

Linear Algebra	
`t()`	transposes a matrix.
`diag()`	generates a diagonal matrix.
`%*%`	matrix multiplication.
`rowsum()`	gives row sums for a matrix.
`colsum()`	gives column sums for a matrix.
`rowMeans()`	gives row means for a matrix.
`colMeans()`	gives column means for a matrix.
`eigen()`	computes eigenvalues and eigenvectors of real or complex matrices.
`svd()`	singular value decomposition of a matrix.
`qr()`	QR decomposition of a matrix.
`determinant()`	determinant of a matrix.
`solve()`	solves linear equations, or computes inverse.

If possible, statistical functions should be used and direct access to the linear algebra functions should be avoided.

Optimisation and Fitting	
`optim()`	general purpose optimisation.
`nlm()`	carries out a minimisation of a function using a Newton-type algorithm.
`lm()`	fits a linear model.
`glm()`	fits a generalised linear model.
`nls()`	determines the non-linear (weighted) least-squares estimates of the parameters of a (possibly non-linear) model.
`approx()`	linear interpolation.
`spline()`	cubic spline interpolation.

Use the online help functions and search for the keyword `smooth` to find more fitting methods.

A.18 Model Descriptions

Mathematically, linear statistical models can be specified by a design matrix X and written generally as

$$\mathtt{Y} = \mathtt{X}\beta + \varepsilon,$$

where the matrix X has to be specified.

R allows us to specify models by giving the rules for how to build the design matrix.

Operator	Syntax	Meaning	Example
$\sim$	$Y \sim M$	Y depends on M	$Y \sim X$ results in $E(Y) = a + bX$
$+$	$M_1 + M_2$	include M_1 and M_2	$Y \sim X + Z$ $E(Y) = a + bX + cZ$
$-$	$M_1 - M_2$	include M_1, but exclude M_2	$Y \sim X - 1$ $E(Y) = bX$
:	$M_1 : M_2$	tensor product, that is, all combinations of levels of M_1 and M_2	
% in %	M_1% in %M_2	modified tensor product	$a + b\%in\%a$ corresponds to $a + a : b$
*	$M_1 * M_2$	"crossed"	$M_1 + M_2$ corresponds to $M_1 + M_2 + M_1 : M_2$
/	M_1/M_2	"nested": $M_1 + M_2$ $\%in\%$ M_1	
^	M^n	M with all "interactions" up to level n	
$I()$	$I(M)$	interpret M; terms in M retain their original meaning; the result determines the model	$Y \sim$ (1 + I(X^2)) corresponds to $E(Y) = a + bX^2$

Table A.57 *Wilkinson-Rogers Notation for Linear Models*

The model specification is also possible for generalised (not linear) models.

Examples:

y ~ 1 + x	corresponds to $\mathtt{y_i} = (\mathtt{1}\ \mathtt{x_i})(\beta_1\ \beta_2)^\top + \varepsilon$
y ~ x	short for y ~ 1 + x (a constant term is assumed implicitly)

y ~ 0 + x	corresponds to $y_i = x_i \cdot \beta + \varepsilon$
log(y) ~ x1 + x2	corresponds to $\log(y_i) = (1\ x_{i1}\ x_{i2})(\beta_1\ \beta_2\ \beta_3)^\top + \varepsilon$ (a constant term is assumed implicitly)
y ~ A	one-way analysis of variance with factor `A`
y ~ A + x	covariance analysis with factor `A` and covariable `x`
*y ~ A * B*	two-factor crossed layout with factors `A` and `B`
y ~ A/B	two-factor hierarchical layout with factor `A` and sub-factor `B`

For an economic transition between models, for example for model comparison, the unction *update()* is available. It updates and (by default) re-fits a model by extracting the call stored in the object, updating the call and evaluating that call, given the new information. In particular, it can be used to re-fit a model to a changed (possibly corrected) data set.

Model Administration	
formula()	extracts a model formula from an object.
terms()	extracts terms of the model formula from an object.
contrasts()	specifies contrasts.
update()	updates and re-fits, or changes a model.
model.matrix()	generates the design matrix for a model.

Example:

```
lm(y ~ poly(x, 4), data = experiment)
```

analyses the data set "experiment" with a linear model for polynomial regression of degree 4.

Standard Analysis	
lm()	linear model. *See also* Chapter 2.
glm()	generalised linear model.
nls()	non-linear least squares.
nlm()	general non-linear minimisation.
update()	update and re-fit, or change a model.
anova()	analysis of variance.

A.19 Graphic Functions

R provides two graphics systems: The basic graphics system of R implements a model that is oriented at pen and paper drawing. The lattice graphics system is an additional second graphics system that is oriented at a viewport/object model. For information about lattice see `help(lattice)`. For a survey about the functions in lattice see `library(help = lattice)`. Information about the basic graphics system follows here. Additional graphics systems are available as packages.

Graphic functions fall essentially in three groups:

"high level" functions.	These define a new output.
"low level" functions.	These modify an existing output.
parametrisations.	These modify the settings of the graphics system.

A.19.1 High-Level Graphics

High-Level Graphics	
`plot()`	generic graphic output.
`pairs()`	pair-wise scatterplots.
`coplot()`	scatterplots, conditioned on covariables.
`qqplot()`	QQ Plot.
`qqnorm()`	Gaussian QQ Plot.
`qqline()`	adds a line to a Gaussian QQ Plot, passing through the first and third quartile.
`hist()`	histogram. See also Section 1.3.2, page 17.
`boxplot()`	box-and-whisker plot.
`dotchart()`	draws a Cleveland dot plot.
`curve()`	evaluates a function or an expression and draws a curve. *Example:* `curve(dnorm, from = -3, to = 3)`
`image()`	colour coded z against x, y.
`contour()`	contour plot of z against x, y.
`persp()`	3D surface.
`matplot()`	plots the columns of one matrix against the columns of another.
`mosaicplot()`	mosaic displays to visualise (standardised) residuals of a log-linear model for the table.
`termplot()`	plots regression terms against their predictors, optionally with standard errors and partial residuals added.

Corresponding function names for the `lattice` graphics are in Table 4.5 (page 143).

A.19.2 Low-Level Graphics

Most high-level functions have an argument `add`. If the function is called with `add = FALSE`, it can be used to add elements to an existing plot. Moreover, there are several low-level functions that suppose that there is already a defined plot environment.

Low-Level Plotting	
`points()`	generic function. Marks points at specified positions. *Syntax:* `points(x, ...)`
`symbols()`	draws symbols at selected points.
`text()`	adds text labels at selected points.
`lines()`	generic function. Joins points at specified positions. *Syntax:* `lines(x, ...)`
`segments()`	adds line segments.
`abline()`	adds a line (in several representations) to a plot. *Syntax:* `abline(a, b, ...)`
`arrows()`	adds a line with arrows to a plot.
`polygon()`	adds polygon with specified vertices.
`rect()`	draws a rectangle.
`axis()`	adds axis.
`rug()`	adds a rug marking the data points.

Besides this, R has rudimentary possibilities for interaction with graphics.

Interactions	
`devAskNewPage()`	controls if a console prompt is given before starting a page of output.
`locator()`	determines the position of mouse clicks. A current graphics display has to be defined before `locator()` is used. *Example:* `plot(runif(19))` `locator(n = 3, type = "l")`
`Sys.sleep()`	suspends execution for a time interval. *Syntax:* `Sys.sleep(⟨seconds⟩)`
`getGraphicsEvent()`	waits for a keyboard or mouse event. Functions to respond to these events can be specified.

(cont.)→

Interactions (cont.)	
	This function needs a graphics display that supports graphics events.

For more interactive facilities, see additional packages, in particular:

- rgl implements OpenGL for real-time 3d rendering,
- rggobi interfaces to the ggobi system for higher-dimensional exploration of data.

A.19.3 Annotations and Legends

The high-level functions generally offer the possibilities to add standard annotations by using arguments:

`main =`	main title, above the plot,
`sub =`	plot caption, below the plot,
`xlab =`	label for the x axis,
`ylab =`	label for the y axis.

For documentation, see `help(plot.default)`.

High-level functions are complemented by low-level functions.

Low Level Annotation		
`title()`	adds main title, analogous to high-level argument main.	
	Syntax:	`title(main = NULL, sub = NULL, xlab = NULL, ylab = NULL, ...)`
`text()`	adds text at specified coordinates.	
	Syntax:	`text(x, y = NULL, text, ...)`
`legend()`	adds a legend block.	
	Syntax:	`legend(x, y = NULL, text, ...)`
	Example:	`plot(runif(100)); legend(locator(1), legend = "You clicked here")`
`mtext()`	adds text to margin.	
	Syntax:	`mtext(text, side = 3, ...)`. The margins are denoted by 1 = bottom, 2 = left, 3 = top, 4 = right).

For annotations, texts some times has to be shortened. Function and variable names can be shortened using `abbreviate()`.

R gives (limited) possibilities for mathematical typesetting. If the text argument is a character string, it is taken directly. If the text argument is an (unevaluated) R expression, R tries to render the expression as usual in a mathematical formula. R expressions can be generated using the functions `expression()` and evalutated with `eval()` or `bquote()`.

Example:

```
text(x, y,  expression(paste(bquote("(", atop(n, x), ")"),
    .(p)\^{}x, .(q)\^{}\n-x\})))
```

`demo(plotmath)` gives several examples for mathematical typesetting in plots.

A.19.4 Graphic Parameters and Layout

Parametrisations	
`par()`	sets parameters for the basic graphics system. *Syntax:* see `help(par)`. *Example:* `par(mfrow = c(m, n))` splits the graphic area in m rows and n columns, to be filled row-wise. `par(mfcol = c(m, n))` fills the area column by column.
`lattice.options()`	sets parameters for the lattice graphics system. *Syntax:* see `help(attice.options)`
`split.screen()`	splits the graphic area in parts. *Syntax:* `split.screen(figs, screen, erase = TRUE)`. If `figs` is a pair of two arguments, these will fix the number of rows and columns. If `figs` is a matrix, each row gives the coordinates of a graphic area in relative coordinates $[0 \ldots 1]$. `split.screen()` can be nested.
`screen()`	selects graphic area for the next graphical output. *Syntax:* `screen(n = cur.screen, new = TRUE)`.
`layout()`	divides the graphic area. This function is not compatible with other layout functions.

A.20 Elementary Statistical Functions

Statistical Functions	
`sum()`	sums up components of a vector.
`cumsum()`	calculates cumulated sums.
`prod()`	multiplies components of a vector.
`cumprod()`	calculates cumulated products.
`length()`	length of an object, for example a vector.
`max()` `min()`	maximum, minimum. See also `pmax, pmin`.
`range()`	minimum and maximum.
`cummax()` `cummin()`	cumulated maximum, minimum.
`quantile()`	sample quantile. For theoretical distributions, use `qxxxx`, for example `qnorm`.
`median()`	median.
`mean()`	mean, including trimmed mean.
`var()`	variance, variance / covariance matrix.
`sort()`	sorting.
`rev()`	reverse sorting.
`order()`	returns a permutation for sorting.
`which.max()`	index of the (first) maximum of a numeric vector.
`which.min()`	index of the (first) minimum of a numeric vector.
`rank()`	rank in a sample.

A.21 Distributions, Random Numbers, Densities...

The base generator for uniform random numbers is administered by *Random*. Several types of generators are available as base generator. **For serious simulation it is strongly recommended to read the recommendations of Marsaglia et al.** (see `help(.Random.seed)`). All non-uniform random number generators are derived from the current base generator. A survey of most important non-uniform random number generators, their distribution functions and their quantiles is given at the end of this section.

R *Random Numbers*	
`.Random.seed`	`.Random.seed` is a global variable that holds the current state of the basic random number generator. This variable can be stored and later be restored with `set.seed()`. Initially, there is no seed. Use `set.seed()` to define a seed. If no seed has been defined, a new one is created based on the current clock time when one is required. Random number generators may use variables other than `.Random.seed` to store their state information. To set a generator to a defined state, always use `set.seed()`. Never set `.Random.seed` directly.
`set.seed()`	initialises the random number generator. *Syntax:* `set.seed(seed, kind = NULL)`

`RNGkind()`	`RNGkind()` gives the name of the current base generator.
	`RNGkind(⟨name⟩)` sets a basic random number generator.
	Syntax: `RNGkind()`
	RNGkind(⟨name⟩)
	Example: `RNGkind("Wichmann-Hill")`
	`RNGkind("Marsaglia-Multicarry")`
	`RNGkind("Super-Duper")`

`sample()`	draws a sample from the values given in vector `x`, with or without replacement (controlled by the value of `replace`).
	Size is by default the length of `x`.
	Optionally, `prob` can be a vector of probabilities for the values of `x`.
	Syntax: `sample(x, size, replace = FALSE, prob)`
	Example: Random permutation:
	`sample(x)`
	
	Biased coin:
	`val<-c("H", "T"")`
	`prob<-c(0.3, 0.7)`
	`sample(val, 10, replace = TRUE, prob)`

If simulations shall be reproducible, the random number generator must be set to a well-defined initial state for a reproduction. So the initial state needs to be recorded. An example is the following statement sequence to store the current state:

```
save.seed <- .Random.seed
save.kind <- RNGkind()
```

These variables can be stored to a file and read from there when necessary. With

```
set.seed(save.seed, save.kind)
```

the state of the random number generator is then restored.

The individual function names for the common non-uniform generators and distribution functions are combined from a prefix and the short name of the distribution (see the list below). General pattern: if `xxxx` is the short name, then

`rxxxx`	generates random numbers
`dxxxx`	density or probability
`pxxxx`	distribution function
`qxxxx`	quantiles

Example:

`x<-runif(100)`	generates 100 random variables with U(0, 1) distribution.
`qf(0.95, 10, 2)`	calculates the 95% quantile of the F(10, 2) distribution.

Distributions	*Short Name*	*Parameter and Default Values*
Beta	`beta`	`shape1, shape2, ncp = 0`
Binomial	`binom`	`size, prob`
Cauchy	`cauchy`	`location = 0, scale = 1`
χ^2	`chisq`	`df, ncp = 0`
Exponential	`exp`	`rate = 1`
F	`f`	`df1, df2 (ncp = 0)`
Gamma	`gamma`	`shape, scale = 1`
Gauss	`norm`	`mean = 0, sd = 1`
Geometric	`geom`	`prob`
Hypergeometric	`hyper`	`m, n, k`
Lognormal	`lnorm`	`meanlog = 0, sdlog = 1`
Logistisch	`logis`	`location = 0, scale = 1`
Negativ-Binomial	`nbinom`	`size, prob`
Poisson	`pois`	`lambda`
Student's t	`t`	`df`

(cont.)→

Distributions	*Short Name*	*Parameter and Default Values*
Tukey Studentised Range	`tukey`	
Uniform	`unif`	`min = 0, max = 1`
Wilcoxon Signed Rank	`signrank`	`n`
Wilcoxon Rank Sum	`wilcox`	`m, n`
Weibull	`weibull`	`shape, scale = 1`

Additional support for generating random numbers is provided by `library(distr)` [40].

A.22 Computing on the Language

The language elements of R are objects, as are data or functions. They can be read and changed like any other data or functions. Chapter 6 of the "R Language Definition" [33] gives details for computing on the language. See also Section 2.1.5, "Function objects" of [33].

Conversions	
`parse()`	converts input into a list of R expressions. `parse` executes the parse, but does not evaluate the expression.
`deparse()`	converts an R expression given in internal representation into a character string.
`expression()`	generates an R expression in internal representations. *Example:* `integrate <- expression(integral(fun, lims))` *See also* 1.3.1: mathematical typesetting in plot annotations
`substitute()`	R expression with evaluation of all defined terms.
`bquote()`	R expression with selective evaluation. Terms in `.()` are evaluated. *Examples:* `n<-10; bquote( n^2 == .(n*n))`

Evaluation	
`eval()`	evaluates an expression.

References

[1] Daniel Adler and Duncan Murdoch. *rgl: 3D Visualization Device System (OpenGL)*, 2008. R package version 0.81.

[2] Richard A. Becker, John M. Chambers, and Allan R. Wilks. *The New S Language*. Computer Science Series. Wadsworth & Brooks/Cole, Pacific Grove, California, 1988.

[3] Ingwer Borg and Patrick J. F. Groenen. *Modern Multidimensional Scaling: Theory and Applications*. Springer Series in Statistics. Springer, New York, second edition, 2005.

[4] John M. Chambers. *Programming with Data*. Springer, New York, 1998.

[5] John M. Chambers and Trevor J. Hastie. *Statistical Models in S*. Chapman & Hall, London, 1992.

[6] William S. Cleveland. *Visualizing Data*. Hobart Press, Summit NJ, 1993.

[7] David B. Dahl and contributions from many others. *xtable: Export Tables to LaTeX or HTML*, 2008. R package version 1.5-3.

[8] Sundar Dorai-Raj. *binom: Binomial Confidence Intervals for Several Parameterizations*. R package version 1.0-3.

[9] James Durbin. *Distribution Theory for Tests based on Sample Distribution Functions*, volume 2 of *Reg. Conf. Ser. Appl. Math.* SIAM, Philadelphia, 1971.

[10] George W. Furnas and Andreas Buja. Prosection views: Dimensional inference through sections and projections. *J. Comput. Graph. Statist.*, 3(4):323–385, 1994.

[11] Peter Gänßler and Winfried Stute. *Wahrscheinlichkeitstheorie*. Springer, Heidelberg, 1977.

[12] Robert Gentleman and Ross Ihaka. Lexical scope and statistical computing. *Journal of Computational and Graphical Statistics*, 9:491–508, 2000.

[13] Gabor Grothendieck and Thomas Petzoldt. R help desk: Date and time classes in R. *R News*, 4(1):29–32, June 2004.

[14] Trevor Hastie, Robert Tibshirani, and Jerome Friedman. *The Elements of Statistical Learning*. Springer Series in Statistics. Springer, Heidelberg, 2001.

[15] Torsten Hothorn, Frank Bretz, and Peter Westfall. Simultaneous inference in general parametric models. *Biometrical Journal*, 50(3):346–363, 2008.

[16] Rob J. Hyndman, David M. Bashtannyk, and Gary K. Grunwald. Estimating and visualizing conditional densities. *J. Comput. Graph. Statist.*, 5(4):315–336, 1996.

[17] Alfred Inselberg. *Parallel Coordinates: Visual Multidimensional Geometry and Its Applications*. Springer, Heidelberg, 2009.

[18] Alfred Inselberg, Tuval Chomut, and Mordechai Reif. Convexity algorithms in parallel coordinates. *J. Assoc. Comput. Mach.*, 34(4):765–801, 1987.

[19] David James and Kurt Hornik. *chron: Chronological Objects Which Can Handle Dates and Times*, 2008. R package version 2.3-24. S original by David James, R port by Kurt Hornik.

[20] Norman L. Johnson and Samuel Kotz. *Discrete Distributions.* Wiley, New York, 1970.

[21] Roger W. Johnson. Fitting percentage of body fat to simple body measurements. *Journal of Statistics Education*, 4(1), 1996.

[22] Bent Jørgensen. *The Theory of Linear Models.* Chapman & Hall, New York, 1993.

[23] Duncan Temple Lang, Debby Swayne, Hadley Wickham, and Michael Lawrence. *rggobi: Interface between R and GGobi*, 2008. R package version 2.1.10.

[24] Friedrich Leisch. Sweave: Dynamic generation of statistical reports using literate data analysis. In Wolfgang Härdle and Bernd Rönz, editors, *Compstat 2002 — Proceedings in Computational Statistics*, pages 575–580. Physica Verlag, Heidelberg, 2002. ISBN 3-7908-1517-9.

[25] Uwe Ligges and John Fox. R help desk: How can I avoid this loop or make it faster? *R News*, 8(1):46–50, May 2008.

[26] Catherine Loader. *locfit: Local Regression, Likelihood and Density Estimation.*, 2007. R package version 1.5-4.

[27] Paul Massart. The tight constant in the Dvoretzky-Kiefer-Wolfowitz inequality. *The Annals of Probability*, 18(3):1269–1283, July 1990.

[28] Rupert G. Miller. *Simultaneous Statistical Inference.* Springer, New York, 1981.

[29] Paul Murrell. *R Graphics.* Chapman & Hall/CRC, Boca Raton, Fla., 2006.

[30] R Development Core Team. *An Introduction to R*, 2008.

[31] R Development Core Team. *R: A Language and Environment for Statistical Computing.* R Foundation for Statistical Computing, Vienna, Austria, 2008. ISBN 3-900051-07-0.

[32] R Development Core Team. *R Data Import/Export*, 2008.

[33] R Development Core Team. *The R language definition*, 2008.

[34] R Development Core Team. *The R Reference Index*, 2008.

[35] R Development Core Team. *Writing R Extensions*, 2008.

[36] C. Radhakrishna Rao. *Linear Statistical Inference and Its Applications.* Wiley, New York, second edition, 1973.

[37] G. M. Reaven and R. G. Miller. An attempt to define the nature of chemical diabetes using a multidimensional analysis. *Diabetologia*, 16:17–24, 1979.

[38] Rolf-Dieter Reiss and Michael Thomas. *Statistical Analysis of Extreme Values.* Birkhäuser, Basel, third edition, 2007.

[39] Brian D. Ripley. *Stochastic Simulation.* Wiley Series in Probability and Statistics. Wiley-Interscience [John Wiley & Sons], Hoboken, NJ, 2006. Reprint of the 1987 original, Wiley-Interscience Paperback Series.

[40] Peter Ruckdeschel, Matthias Kohl, Thomas Stabla, and Florian Camphausen. S4 classes for distributions. *R News*, 6(2):2–6, May 2006.

[41] Deepayan Sarkar. *lattice: Lattice Graphics*, 2008. R package version 0.17-15.

[42] Günther Sawitzki. Numerical reliability of data analysis systems. *Computational Statistics & Data Analysis*, 18(2):269–286, 1994.

[43] Günther Sawitzki. Report on the numerical reliability of data analysis systems. *Computational Statistics & Data Analysis*, 18(2):289–301, 1994.

[44] Günther Sawitzki. Quality control and early diagnostics for cDNA microarrays. *R News*, 2(1):6–10, March 2002.

[45] Günther Sawitzki. *lshorth: The Length of the Shorth*, 2008. R package version 0.1-4.

[46] Galen R. Shorack and Jon A. Wellner. *Empirical Processes with Applications to Statistics.* Wiley, New York, 1986.

[47] W. S. Stiles and J.M. Burch. NPL colour-matching investigation: Final report. *Optica Acta*, 6(1-26), 1959.

[48] Antony Unwin, Martin Theus, and Heike Hofmann. *Graphics of Large Datasets: Visualizing a Million.* Springer, Heidelberg, 2006.

[49] Thomas Lumley using Fortran code by Alan Miller. *leaps: regression subset selection.* R package version 2.7.

[50] Paul Velleman. *Data Desk Handbook, Statistics Guide, Reference Guide, and Quickstart Guide.* Data Description, Ithaca, NY, sixth, revised and extended edition, 1997.

[51] William N. Venables and Brian D. Ripley. *S Programming.* Statistics and Computing. Springer, New York, 2000.

[52] William N. Venables and Brian D. Ripley. *Modern Applied Statistics with S.* Springer, Heidelberg, fourth edition, 2002.

[53] John Verzani. *UsingR: Data sets for the text "Using R for Introductory Statistics"*, 2007. R package version 0.1-10.

[54] Ron Wehrens and Bjørn-Helge Mevik. *pls: Partial Least Squares Regression (PLSR) and Principal Component Regression (PCR)*, 2007. R package version 2.1-0.

[55] Sanford Weisberg. *Applied Linear Regression.* Wiley Series in Probability and Statistics. Wiley, New York, third edition, 2005.

[56] G. N. Wilkinson and C.E. Rogers. Symbolic description of factorial models for analysis of variance. *Journal of the Royal Statistical Society Series C*, 22:392–399, 1973.

[57] Samuel S. Wilks. *Mathematical Statistics.* John Wiley & Sons, New York, 1962.

Functions and Variables by Topic

Graphics

Add to Existing Plot / Internal Plot

Computations Related to Plotting

High-Level Plots

Interacting with Plots

Colour, Palettes etc

Dynamic Graphics

Graphical Devices

Basics

Basic System Variables

Environments, Scoping, Packages

Data Manipulation

Data Attributes

Data Types (not OO)

Character Data (String) Operations

Categorical Data

Missing Values

Lists

Dates and Times

Mathematics

Matrices and Arrays

Linear Algebra

Basic Arithmetic and Sorting

Mathematical Calculus etc.

Logical Operators

Optimisation

Programming, Input/Ouput, and Miscellaneous

Programming

Files

Connections

Looping and Iteration

Methods and Generic Functions

Printing

Error Handling

Session Environment

Utilities

Miscellaneous

Documentation

Debugging Tools

Statistics

Probability Distributions and Random Numbers

Simple Univariate Statistics

Statistical Inference

Statistical Models

Regression

Non-linear Regression

Multivariate Techniques

Curve (and Surface) Smoothing

Function and Variable Index

Subject Index